The Copenhagen Conspiracy

The Copenhagen Conspiracy

David Ferry

PAN STANFORD PUBLISHING

Published by

Pan Stanford Publishing Pte. Ltd.
Penthouse Level, Suntec Tower 3
8 Temasek Boulevard
Singapore 038988

Email: editorial@panstanford.com
Web: www.panstanford.com

British Library Cataloguing-in-Publication Data
A catalogue record for this book is available from the British Library.

The Copenhagen Conspiracy

ISBN 978-981-4774-75-8 (Hardcover)
ISBN 978-1-351-20723-2 (eBook)

Contents

Preface

As the nineteenth century was drawing to a close, there occurred a rash of technological breakthroughs that would ultimately presage the information revolution a century later. There was Bell's telephone in the 1870s, Edison's carbon microphone and incandescent light bulb just before 1880, development of early automobiles in the 1880s, Marconi's wireless telegraph in the 1890s, and many early experiments in flight which led to the Wright brothers in 1903. In fact, technology had progressed so rapidly that Albert Michelson, perhaps the preeminent American scientist of the period, is said to have remarked in 1894 that the grand underlying theories of science have been firmly established and the future truths of science are to be looked for in small changes to known values. It was even suggested to close the Patent Office. Yet, at the close of the nineteenth century, we stood on the threshold of one of the greatest periods of science, in which the entire world and understanding of science would be shaken to the core and greatly modified. This explosion of knowledge led ultimately to that same information revolution that we live in today. Planck and Einstein showed that light was not continuous, but made of small corpuscles that today we call photons. Einstein changed the understanding of mechanics with relativity theory, airplanes became conceivable, radio and television blossomed, and the microelectronics industry, which drives most of modern technology, came into being. New areas of science were greatly expanded and developed, and one of these was quantum mechanics, which is the story to be told here.

What is interesting is that quantum mechanics may well be the last area of physical science that clings to the nineteenth-century views of Ernst Mach. Most people know of him from the Mach number, the speed of an aircraft relative to the velocity of sound. Fewer know that he was the de facto leader of the Viennese philosophy community in the latter parts of the nineteenth and early twentieth century. One can perhaps summarize his philosophy by saying that he did not believe in atoms because he couldn't see them. This view stemmed from a belief that reality was contingent on actual observations or conscious awareness of the event. It is perhaps best described by the elementary school puzzle as to whether a tree, falling in the forest, makes a sound if no one is there to hear it. Perhaps it is worth considering this. One could of course put a microphone in the forest to record any events that happened. Then, does the reality of the tree arrive only when we listen to the recording. Or, indeed, does the recording only become real when we listen to it. I suspect that rational people would decide otherwise. Yet, in quantum mechanics, this is still the accepted interpretation of the field.

It is important to understand that I am using Mach as a symbol of what has been called by some as "positivism," although it may better be referred to as "negativism" when we discuss the theoretical foundations of modern physics. I am also fully aware that there were a great number of philosophers of science who contributed to this, as well as many more modern ones who have modified it. Nevertheless, I will still use Mach as the paradigm. The reader will find citations to Philipp Frank as well, and I am fully aware that he was one of the founders of the Vienna Circle, a group of Austrian philosophers and physicists, who often were more Machist than Mach himself. It also is true that a great percentage of scientists at the start of the twentieth century actually were infected with the Mach point of view. There certainly was an ongoing dispute between the positivists and the more realistic views of many physicists. The rise of quantum mechanics occurred in the midst of these disputes. Einstein himself is said to have been a devotee when young. But he rapidly realized that his view of reality could be explained and advanced with the development of true theory, such as relativity, particularly special relativity. In Mach's view, such a theory, without experimental observation, could not be a description of reality. Yet, Einstein's curved space and gravity waves, Dirac's positrons, and even Higg's boson were all accepted by most of science before the confirming experiments were ever performed. It's true that Dirac's prediction lay in the field of quantum mechanics, but it was not accepted by some until the experimental observation of the positron was made, and they could no longer push it aside.

As remarked above, science and technology progressed tremendously in the 1880s. In technology, this led to an economic boom, but, as with many such booms, it became overextended. Financial problems in Argentina (what's new?) and the failure of several American railroads led to the financial crash of 1893. In science, the success of Boltzmann and statistical physics was met with the antagonism of Mach and his followers, leading to what I will call the religion of positivism. This continued to grow into the new century with the founding of the Vienna circle, and similar efforts elsewhere. Perhaps this rise should be called the intellectual crash of the 1890s.

The high priest of this religion in quantum mechanics was Neils Bohr, and his viewpoint could be reinforced as most young scientists of the period felt a need to spend time in Copenhagen. He was a leader, along with his disciples, in the formulation of quantum mechanics and thus built this view into the underlying philosophy of quantum mechanics. Thus the religion was passed on to each new disciple and bred into the leadership of the coming quantum revolution. One view is, "Here I only note that those who accept Bohr's way of thinking speak of the spirit of Copenhagen ... while those who are critical speak of the Copenhagen orthodoxy."[a] It was not all Bohr, but also sermons of his disciples who had passed through Copenhagen: "he [Bohr] must be approached as a physicist and not as a philosopher.... The problem is ... that statements ... of the standard interpretation or the Copenhagen interpretation of quantum mechanics are often very far off the mark ... [and] often attribute to Bohr conceptions that would more properly reflect those of Werner Heisenberg, Wolfgang Pauli, or John von Neumann."[b] Nevertheless, these were all affected by an intense subjugation to Bohr's pressure, which some have viewed, using a post-WW2 phrase, as brainwashing.

---

[a] A. Pais, *Niels Bohr's Times, in Physics, Philosophy, and Polity* (Clarendon Press, Oxford, 1991) p. 24.

[b] P. McEvoy, *Niels Bohr: Reflections on Subject and Object* (MicroAnalytix, San Francisco, 2000), p. 9.

There are some who would say that Bohr set back the real understanding of quantum mechanics by half a century. I believe they underestimate his role, and it may be something more like a full century. Whether we call it the Copenhagen interpretation or the Copenhagen *orthodoxy*, it is the *how* for the continuing mysticism provided by Mach that is still remaining in quantum mechanics. It is not the *why*. *Why* it perseveres and *why* it was forced on the field in the first place is an important perception to be studied.

In this book, I want to trace the development of quantum mechanics and try to uncover the *why*. Any strong exposition of a viewpoint is often termed philosophy, but philosophy is the study of knowledge—what can be known and what is known within these limits. I am not a philosopher, but I do have strong opinions about these questions as they pertain to quantum mechanics. To me, the story involves the same factors that are present in modern scientific controversies—ego, recognition, status, and the desire to suppress any views that negatively affect these factors (and, yes, there is some role to be played by sex). The early pioneers certainly had, and encountered, all of this. In fact, it is unfortunate that science has not evolved from these earlier frailties, because these same factors still impact modern science—one only has to look at global warming to see that this is still true.

Over the years, I have argued with a great many of my colleagues over this subject. But, like proper religion, this theological discussion leads to no agreements or conclusions. Rather, at times it seems more like tribal apes asserting leadership by beating upon their chests, much as ancient Vikings beat sword upon shield prior to major battles if for no other reason than to demonstrate to the other side their willingness to kill or be killed. Fortunately, these theological discussions have not come quite so far, at least up until the present day. Some of these colleagues, with whom I have argued, profess the standard Copenhagen orthodoxy, while others want to go further into more radical interpretations. So, in the past few years, I have resolved to write down my view of history and my arguments for various interpretations. As for many of my projects, Stanford Chong of Pan Stanford Publishing encouraged me to go ahead with the project. I would especially like to thank Jon Bird, Karl Hess, Alex Kirk, Lavinia Sebastian, Carlo Requiao da Cunha,

Josef Weinbub, Mihail Nedjalkov, James Huffman, and Laszlo Kish, who have all suffered through a reading of the drafts of this effort. They have made a great many suggestions to clarify my not always understandable logic and often suggested ways to make a point more clearly. Of course, they have not always succeeded, and any remaining confusion is just mine alone.

Chapter 1

In the Beginning

As he rose from his seat and moved toward the podium, he was nervous. It was not too cold in Berlin, which was unusual for this time of year—14 December 1900 [1]—so perhaps it was the warmth in the room, or perhaps it was the revolution he was preparing to launch. He had talked about his work on black-body radiation previously, the last time just the previous October. But now the work was basically complete and he had to convince the audience that the assumption he had been forced to make was justifiable on the grounds of physics. At an earlier meeting, colleagues from the same institute had presented new data on black-body radiation, covering a much wider part of the spectrum. In the intervening months, Max Planck had developed a newer and more complete version of his theory, but in doing so had to drop the view held in classical physics that light was a form of electromagnetic radiation which consisted of a continuous wave field, much like the waves we think of in a body of water. Instead, Planck felt forced to introduce a radical new idea, in which radiation, such as light, was composed of corpuscles of energy, whose value was related to their frequency (or color). The proportionality constant he introduced is today known as Planck's constant. It was this new theory he was preparing to talk about at

The Copenhagen Conspiracy
David Ferry
Copyright © 2019 Pan Stanford Publishing Pte. Ltd.
ISBN 978-981-4774-75-8 (Hardcover), 978-1-351-20723-2 (eBook)
www.panstanford.com

this meeting of the Deutsches Physikalische Gesellschaft (German Physical Society).

At the time, Planck had been a full professor in Berlin for almost a decade. In coming to Berlin, he had succeeded the famous Gustav Kirchoff, which was a coup of sorts. And he had recently succeeded in convincing the various local physical societies in Germany to merge into a single society. So, he was held in high regard within the physics community. Originally, he was born in Kiel, the capital of Holstein, in 1858. At the time Holstein was part of Denmark but also a member of the German confederation—it would be united with Germany a decade later as Prussia united the German states. In 1867, his family moved to Munich, and Planck went to the University of Munich, where he eventually concentrated on theoretical physics. As Planck himself said, "I studied experimental physics and mathematics; there were no classes in theoretical physics as yet" [2]. Prior to his graduation, he went to Berlin for a year to study with Kirchoff and Helmholtz, as well as with the mathematician Weierstrass. He finally graduated in 1879, but it was not until 1885 that he was offered a faculty position at the University of Kiel. The world had turned full circle on him. Then, in 1889, he was summoned to Berlin to replace Kirchoff. Throughout his career, he worked on thermodynamics and would turn to statistical physics in Berlin. In Berlin, he had many colleagues, particularly in experimental physics, and this would lead to his success he chose to present that December evening.

In truth, the idea that light was corpuscular in nature can be traced to the ancient Greeks. It is even discussed in that manner by Newton in his treatise on optics in the 17th century. But later investigations, particularly by Thomas Young at the start of the 19th century, had come to the conclusion that light was composed of continuous waves, and this fit nicely into Maxwell's theory of electricity and magnetism a few decades later. Indeed, Young's observations of interference patterns arising from passing light through two closely spaced holes in a blocking panel, clearly demonstrated the wave-like nature of light in 1803 [3]. Most of the studies of light and optics expanded upon this wave theme throughout the 18th century. Indeed, Kirchhoff, to whose chair Planck ascended at Berlin University upon the former's death, had demonstrated laws of diffraction in his own thesis, which also demonstrated the wave nature of light. To now reintroduce the corpuscular nature of light would be to turn back two centuries of physical thought.

The new theory was basically a statistical theory, since it would include a probability distribution function, one which ultimately would be known as the Bose–Einstein distribution. In his development, there actually were two new constants, which he believed were fundamental. The first, which related temperature to energy, he named the Boltzmann constant, after Ludwig Boltzmann, an Austrian who had been a major player in developing statistical physics and the probabilistic approach. Planck was not initially disposed toward the statistical theory, but had come to appreciate it, particularly with the work of Boltzmann. Indeed, he had to use the statistical concepts to develop his new theory, so he felt it appropriate to name the new constant for him. The second new constant, already mentioned, related the frequency (or color) of light to the energy of the corpuscle of light through the relation $E = hf$, where E is the energy, f is the frequency, and h was the new constant he established. An interesting aspect was that, without the aid of modern computers (or any computer for that matter) or calculators, he had determined the values within 1–2% of today's accepted values for these constants. This was an amazing achievement.

Apparently, the talk was well received and, after the manuscript was finally published [4], Planck found himself the subject of widespread discussion over his revolutionary ideas. But he was somewhat reluctant about this, because it was after all revolutionary, and still only a theory. To be sure, his results fit well with experimental data on the radiation, but the quantization of the light remained only a hypothesis, yet to be verified. The units on his new constant are energy times time, which physicists refer to as *action*. He himself referred to this new constant as the constant of action. As he remarked later, he knew at the time that

> either the quantum of action was a fictional quantity, then the whole deduction of the radiation law was in the main illusory and represented nothing more than an empty non-significant play on formulae, or the derivation of the radiation law was based on

a sound physical conception. In this case the quantum of action must play a fundamental role in physics, and here was something entirely new, never before heard of, which seemed called upon to basically revise all our physical thinking, built as this was ... upon the acceptance of the continuity of all causative connections [5].

Even at the time of his Nobel speech, the full form of the revolutionary development following from his paper was still not apparent, but his quantization of light had been verified.

Black-Body Radiation

If we set an object outside in the sun, especially in Arizona, it gets hot. Our object absorbs energy from the sun and, in order to establish an equilibrium, it re-radiates a large amount of that energy. In doing so, the temperature rise is necessary in order to re-radiate as much energy as it absorbs from the sun. This re-emission of the energy is termed black-body radiation, and the amount of energy that is radiated is proportional to its temperature. Now, the human body has an average temperature around 98.6°F, which is 37°C, or 310.15 K (the Kelvin value is equivalent to °C, but referenced to absolute zero, which is −273.15°C). So, a person radiates energy corresponding to that temperature, but none of this is in the visible regime, as the frequency is just too low. But we feel it as body heat of a nearby person. We all have been in a room that had too many people and too little airflow so that the temperature rise was uncomfortable. This results from the combined effect of each person's radiated energy. The sun, on the other hand, has a surface temperature of nearly 5800 K, so a significant fraction of its emission lies in the visible spectrum and corresponds to the light we perceive. This much has been known for as long as people have lived together or scientists have observed the sun and the stars. But scientific interest accelerated in the late 1870s following Thomas Edison's development of the long-lasting electric light bulb. Now, Edison didn't invent the concept, but he developed the technology to make it have a meaningful lifetime, which gave birth to the industry. In this light bulb, the passage of a current through a thin filamentary wire, causes it to become heated to such a temperature that visible light emission is observed. This principle continues today in each and every incandescent light bulb we use. The fact that only a small fraction of the radiation lies in the visible spectrum tells us why these bulbs are so inefficient. Most of the energy radiated is useless for vision and appears only as heat, since the black-body radiation spectrum is so broad.

The real attempts to try to explain the spectrum of black-body radiation are believed to have begun with Ludwig Boltzmann, whom we met earlier. In the 1880s, he began to develop theoretically an idea on the emission using the statistical physics which he favored. He found a relationship for the total amount of energy that could be radiated, which was a functional form depending upon the temperature of the body, raised to the fourth power. Most importantly, he developed a probability function which has become very important in classical physics, but it was Wilhelm Wien (who published under Willy Wien) that developed the best classical formula for black-body radiation. Beginning from earlier work by von Lommel and by Michelson, he came to the result that the intensity (power per unit area of the radiation) could be expressed by the formula [6]

$$I(\lambda) = \frac{C}{\lambda^5} e^{-a/\lambda T},\tag{1.1}$$

where C and a are constants, λ is the wavelength of the light, and T is the absolute temperature (in degrees Kelvin). Rewritten in terms of the frequency (which is c/λ, where c is the speed of light), this has become known as Wien's law. In his paper, Wien acknowledged that Paschen had arrived at a very similar formula but had not pinned down the value 5 for the exponent in the pre-factor. So, when experimental confirmation arrived that fit this expression almost perfectly, it was Wien who was awarded the Nobel prize (in 1911). An important aspect of Wien's law is that it predicts a peak in the black-body radiation, which is important as this peak can be equated to particular temperature characteristic of the peak. It is remarkable that Wien had made such good progress only to learn that he had missed the important quantization of light brought about by Planck only a few years later. Nevertheless, Wien was awarded the 1911 Nobel Prize, even though it was known by then that his formula was wrong. Others continued to work in this area, and in particular

Rayleigh [7] and Jeans [8], who neither captured the exponential decay nor got the proper value of the exponent in the pre-factor, as they found the power 4 rather than 5. As we would say today, they were too little too late.

So, it appears that at the end of the 19th century, the problem of black-body radiation had become mainstream, with a wide range of scientists working at trying to explain the experimentally measured spectrum. Planck himself had begun working on the problem already in 1890. His efforts thus spread across a decade of effort, but he was aided by various groups in Berlin, who were conducting experiments on the subject. As a result, he gained access to data sooner than the rest of the world. Even Wien law's seemed to fit pretty well, but as the end of the century approached, deviations from Wien's law were beginning to appear. The first of the problems came from H. Beckmann, a student at the university in Tübingen, who was studying the radiation [9]. Two groups immediately began to work to confirm the deviations reported by Beckmann and were successful in this. These were Otto Lummer and Ernst Pringsheim at the Physikalisch-Technische Reichsanstalt, the predecessor of today's German national bureau of standards, in Charlottenburg, a suburb of Berlin, and Heinrich Rubens and Ferdinand Kurlbaum at the Technical University in Charlottenburg. The first results, from Lummer and Pringsheim, confirmed Beckman's results [10, 11]. They plotted the observed radiation as a function of wavelength for several temperatures and found a deviation from Wien's law at long wavelength. This deviation became worse at higher temperatures. Then, Rubens and Kurlbaum plotted the radiation from a fixed wavelength as a function of temperature [12], showing that Beckmann, Lummer, and Pringsheim's and their own data all agreed. Also, they compared the results with Wien's formula and showed a large deviation. These results were first presented at the October 1900 meeting of the Prussian Academy of Science, where Planck also presented his preliminary results. They obviously discussed this, as their journal paper also contains the fit from Planck's formula, which was almost perfect [13]. These studies confirmed the work of Beckmann and showed that Wien's formula was not general, and significant deviations could be observed. What they had found was that the black-body radiation increased linearly with temperature, rather than exponentially as shown in Eq. 1.1, at least for the temperatures and spectral ranges that they could study. As mentioned, it was useful that both of these groups were in Berlin, as it allowed Planck to access to the data prior to its publication, and they benefitted from his theory. Thus, Planck was able to formulate his new law in a generic form, which would finally be realized in December as

$$I = \frac{2hf^2}{c^2} \frac{1}{e^{hf/kT} - 1}, \tag{1.2}$$

where, to reiterate, f $(= c/\lambda)$ is the frequency (color) of the light, c is the velocity of light, h is Planck's constant, and k is Boltzmann's constant. It is clear that he has replaced the Boltzmann distribution in Wien's law with a new distribution (second term) that fits to the data. But, in his October version, neither h nor k were present, only adjustable constants such as those in Wien's law. It was his final recognition that these new constants were necessary, but were also very fundamental in nature, which led to his conclusions. The second of these, k, could be determined by measurements of the entropy, a thermodynamic quantity which came from the measurements. But it was the first of these, h, which would create the revolution. A key factor is also that, for small values of the photon energy, relative to the thermal energy, the exponential can be expanded, and the intensity becomes linear in temperature as the recent experiments had shown.

Now, if one were content to just measure the energy of the light rays that came from the experiment, the quantization of light would not be so noticeable, but the above equations are in terms of the intensity of light, that is, the power per unit area that is in the radiation[a]. Power is so much energy per unit time, and quantization of the energy meant that the intensity could not be continuous. Rather it measured, on average, so many corpuscles of light per second striking the surface, and this had to be an integer number. So, the intensity could not increase continuously, but must be discontinuous, increasing by one or more units of energy. This is a

---

[a]There are a variety of strict definitions of intensity, some involving the spectral width being measured. Here, we use the equivalent to irradiance as the W/m^2 of incident energy, where a watt is a joule per second.

significant break from the wave picture given by Maxwell's theory of electricity and magnetism. Today, we take this in stride as modern electronics has created detectors which are capable of measuring the arrival of a single corpuscle of light.

The black-body radiation law is very important to modern electronics. For example, the details of the emission spectrum of "light" from the sun is very important for creating photovoltaic devices which optimize the conversion of sunlight to electrical energy. Hence, the details of the selection of various materials and the structure optimization are dependent upon the exact details of the solar spectrum after it has passed through the earth's atmosphere. A body at room temperature, which is not illuminated by another source, appears black, because its radiation is entirely in the infrared region—that region whose wavelength is larger than that of visible radiation (and whose corpuscle energy is smaller than that of a visible corpuscle). Nevertheless, this infrared radiation is quite useful for a variety of applications, not the least of which are so-called night vision image convertors. If we use infrared detectors, a person moving among shrubbery is quite visible as his or her emission intensity is much larger than that of the surrounding shrubbery. Vehicles with their quite hot engines are even more visible. As a result, infrared image convertors are extremely valuable in a large range of commercial, as well as military, applications. In Fig. 1.1, we illustrate the intensity of this black-body radiation for three sources: one at the human body temperature (300 K), one at 3500 K, which is close to that of the incandescent lamp, and one at 5500 K, which is close to the surface temperature of the sun as well as to some of the modern "white light" light-emitting diodes (although these latter devices do not have a "thermal" emission spectrum). One can see here that the highest temperature gives a spectrum which maximizes the light in the visible (it will appear whiter because of the variation in the eye's sensitivity with energy), while the mid-temperature emphasizes the red end of the spectrum, which gives incandescent lamps their orangish nature. And it is clear that the human doesn't emit any radiation in the visible.

The main confirmation of Planck's quantization of light would come from another direction, and from a young scientist whose subsequent career would lead him to be called the century's best

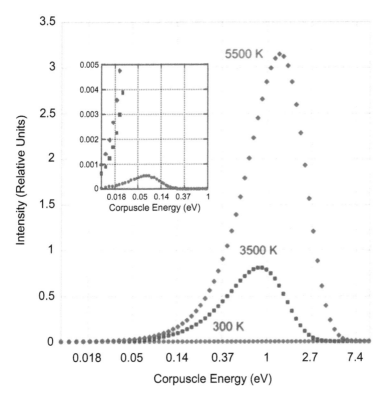

Figure 1.1 The black-body radiation intensity as a function of the natural log of the energy of the corpuscle. The 300 K curve can hardly be seen, but the low energy region is expanded in the inset to show this region better. The visible spectrum is considered to be from about 1.6 to 3.2 eV.

physicist, and perhaps the best physicist of all time, but his results would be followed by many more scientists and experiments, all confirming the quantization of light. Within a decade of Planck's talk, there would be no longer any doubt that his assumption was exactly correct.

The Photoelectric Effect

What we call the photoelectric effect appears to begin with a study by Heinrich Hertz [14] in 1887. To be sure, there were several earlier papers dealing with what we now call the photovoltaic effect. However, Hertz discovered that shining ultraviolet light (light with a wavelength much shorter than that of the visible spectrum, so that the frequency was *beyond violet*, the highest visible frequency) on a metal led to some sort of emission that would make creating a spark, through the breakdown of air, much easier. It was later found by a variety of authors that the emission was electrons coming from the metal. If one negatively charged a metal, then the ultraviolet light would cause this charge to dissipate. Similar effects did not arise if the metal was positively charged. Moreover, the current that would flow from the metal to another electrode would increase with either the frequency or the intensity of the light. Now, the current could increase if the number of electrons increases or if the velocity (energy) of the emitted electrons increased. Just a year later, the Russian Stoletow demonstrated a more efficient experimental setup [15] and eventually it was shown by Lenard [16] that the velocity of the emitted particles was a function of only the frequency of the light and not of its intensity. Now, this created a problem. If light was a wave subject to Maxwell's laws of electricity and magnetism, then one expected the energy of the emitted particles to increase with either the intensity of light or with the frequency of light, and this didn't seem to be the case. So, the classical theory of light had again a major problem with an experiment in which light was the causative source for the observations. Many other scientists performed experiments on the effect, but only confirmed these basic relationships. Increasing the intensity of light increased the number of emitted particles, but did not increase their energy. Another unusual effect was that there was a critical frequency—for light below this critical frequency, no electrons were emitted. First, black-body radiation, and now the photoelectric effect were challenging the nice comfortable world of classical physics. The solution would come from a relatively unknown source.

In 1905, Albert Einstein was a patent examiner in the Swiss patent office in Bern. Although born in Germany, he was educated in Zürich and had become a Swiss citizen in 1901. Even though it was recognized that he was outstanding in math and physics, it was still difficult for him to find a job when he graduated from the Zürich Polytechnic in 1900 with a teaching certificate. Eventually, this led him to the patent office. While he had published a few relatively obscure papers, mostly in thermodynamics, prior to 1905, this year was to be his miracle year. In this year, he not only completed his dissertation for the doctorate (from the University of Zürich) on a method of measuring the size of molecules, he published four papers in disparate fields of physics, each of which revolutionized our understanding of physics. The first of these was an explanation of the photoelectric effect [17], which is our concern here. The second of these was an explanation of what today we call Brownian motion, which is based in statistical physics and probability. This will be important to us later. The third paper established special relativity and changed the laws of motion forever. The fourth paper established the famous relation between mass and energy. Further papers on some of these topics appeared in the following years. To say that the scientific establishment was knocked for a loop by the end of the year would be an understatement. Most scientists hope to make at least one groundbreaking achievement in their career. Many do more, but to achieve four such seminal papers within one year is truly remarkable, especially for one so young. His first paper was received by the journal only a few days after his 26th birthday. But he was no longer relatively unknown! Within a couple of years, he had his choice of prestigious university positions—within 9 years, he was director of the Kaiser Wilhelm Institute for Physics and professor at Humboldt University, both in Berlin. He would remain in Berlin until emigrating to the United States in 1933.

The key which Einstein used to unlock the questions about the photoelectric effect was provided by Max Planck. Once Einstein realized that the light coming in was composed of corpuscles of fixed energy, determined by their frequency through Planck's relation $E = hf$, the rest of the problem unfolded via classical particle dynamics. He assumed that the corpuscle of light, which he called the *lichtquant* (light quantum), would interact with the electrons in the solid, and it is very interesting how he blended quantization of light with classical scattering physics. In the years after this work, the light quantum came to be called the photon, and we will use this term here. In this interaction, the photon is absorbed and transfers its energy and momentum to the electron. To understand how this explains the observations, we have to first address the critical frequency, above which one sees electron emission from the solid.

Every material solid has a potential barrier at the surface which serves to keep the electrons in the material. In metals, the electrons are distant from the atomic nuclei, and thus can be more easily moved around in the material. If this barrier were not present, the electrons would be likely to fall out to the ground. The barrier itself arises from just this effort of the electrons to leave the solid, so that there is a dipole at the surface between the electrons moving just away from the nuclei at the surface and the positive charge of the nuclei pulling them back. This process creates a surface dipole, which serves to create a potential barrier, holding the electrons inside the solid. For obscure reasons, this potential amplitude is called the *work function* of the solid. To understand how this fits into the model, consider that you live in a village located within a great meteor crater, such as the one in northern Arizona. If you wanted to leave, you would have to climb to the edge of the crater to reach the plateau above. Anyone who has been in the mountains understands that it takes a great deal of effort to climb upward, and that the steepness of the climb makes it more difficult. So, to reach the top of the crater wall requires a considerable amount of energy. If you want to move away from the crater once you have scaled the wall, you need additional energy. If you use it all just to climb the crater wall, you will collapse at the top. Moving away from the crater can be achieved only if you start with more energy than that required to climb out of the crater. This is also true of our electrons, which have gained the energy of the photon. In order to leave the solid, they must have more energy than that required to reach the top of the potential barrier. Then, their remaining energy can be used to move away from the solid. Hence, the height of the potential barrier sets the minimum energy required for the photon, and this amount corresponds to the critical frequency through Planck's relation. So, the minimum energy required for electron emission was a property of the solid material being examined and was not a property of the light itself. This was important, as it established that light could be used to examine the properties of solid material. Today, this is an important method for the study of materials, and very intense light sources such as the National Light Source at Brookhaven National

Laboratory are facilities to which scientists go to use light to study materials of their own choice. Here, this light is far beyond our concept of visible light, and these sources provide intense X-rays over a wide range of energies.

Light, which is an electromagnetic wave as shown by Maxwell much earlier, has two principle properties. First, there is the frequency of light, which Planck showed determined the energy of the photons through his relation. As the frequency is increased, the excess energy the particle has after escaping the solid determines its velocity. Higher excess energy means higher velocity. The current depends upon both the number of electrons that escape and their velocity (in steady state, the current is the product of these two quantities, the electron charge, and some geometrical factors). So, increasing the frequency means increasing the velocity, which in turn means increasing the current, and this agrees with the experimental observables. The second property is the intensity of light. In classical physics, it was thought that an increase in intensity would correlate to an increase in the energy of the electrons that escaped. But this did not turn out to be the case. Once again, Planck's quantization comes into play. When light is quantized, the intensity tells us the *number* of photons that reach the solid. One photon can yield at most one electron escaping from the solid. Hence, increasing the intensity did not affect the energy of the escaping electrons, but did increase the number of such electrons. Since the number of electrons increases, so also does the current, again in agreement with the experimental observables. Thus, Einstein could explain the photoelectric effect in a very straightforward manner once Planck's quantization of light was put into the problem.

But this is the view looking back on the problem. Einstein was looking at more than just solids, and in fact had a deep interest in the ionization of gasses. So, he began by re-examining the black-body radiation problem as an approach to assure that all of his readers would so to speak "be on the same page." This meant a detailed discussion of energy and entropy, consistent with approaching the problem from the view of thermodynamics, a field with which he was familiar, and in which he had earlier published papers. In the end, however, it came down to obtaining exactly the results that are discussed above. What is remarkable about the so-called scientific leadership is that it was this work which earned him the Nobel Prize, and not the theory of relativity which made him famous far beyond the scientific world.

Further Confirmation

In addition to the photoelectric effect, Einstein also showed how Planck's quantization was important in what is known as the specific heat of a body. Since the development of thermodynamics in the 19th century, it was realized that heat was stored in a solid (or a gas) in the motion of the atoms. The heat capacity became a measure of this heat storage for a given material, while the specific heat was the variation of the heat capacity with the mass of the solid. Planck had already shown in his black-body radiation theory that the energy associated with heat could be described in terms of a series of oscillators, which represented the different modes by which the coupled motion of the atoms could be described. There were as many modes as there were atoms. Einstein connected the fact that the quantization of the radiation from the solid was connected to a quantization of the vibrational energy of each of these oscillators, and this led to the ability to calculate the specific heat for the solid. His work was extended by several authors, including Max Born, Theodor von Karman, and Peter Debye. These efforts were corroborated by measurements done by Walter Nernst. Nernst also applied the quantization of light in dealing with the specific heat of gasses. But it was the light quantum which was important in all these studies. Max Born will play a continuing role in the rise of quantum mechanics.

An important breakthrough was achieved by Gustav Hertz and James Franck in 1911, in their studies of inelastic collisions of electrons with the atoms in a gas [18]. In this experiment, they showed that an incident electron could cause a gas atom to emit light. First, there was a critical amount of energy that the electron needed before light could be emitted. Then, the amount of light was proportional to the number of electrons in the excitation beam. In essence, this is the inverse of the photoelectric effect. Moreover, they showed that the light that was emitted had only specific energies, which was significant in its own right. Later, they would connect this to Bohr's model of the atom, which we discuss in the next chapter.

In 1915, William Duane and Franklin Hunt studied the generation of X-rays by accelerating electrons to high voltage and then impacting them on a tungsten target [19]. Such an effect produces characteristic X-rays corresponding to definite transitions in the atoms of tungsten, but also leads to a broad background radiation, which is thought to arise from bremsstrahlung emission of the electrons inside the tungsten (this is emission that comes from the acceleration of the electrons, an acceleration that is required to maintain their circular orbits). In measuring this broad spectrum, they discovered that there is a maximum frequency, or minimum wavelength, for the emitted X-rays which is governed by Planck's quantization of light. If the maximum energy that can be transferred to the X-ray photon is that of the accelerating potential of the electrons, then this gives the maximum possible energy of the photon. The frequency of this maximum energy photon is given by Planck's relation, and this allows a new method of determining the value for Planck's constant. This relationship is now known as the Duane–Hunt law in X-ray spectroscopy. Their results were confirmed in experiments by David Webster [20].

Emil Warburg was studying the onset of chemical reactions under illumination, and found that there was a critical minimum frequency of light that was required to initiate the reaction [21]. He connected this discovery with the photoelectric effect discussed by Einstein, with whom he was acquainted. Taking some detailed measurements, he was also able to agree on the value for Planck's constant.

So, within a decade of Einstein's photoelectric effect paper, Planck's quantum of action had been confirmed across a wide range of physics and chemistry. It was clear that the assumption of absolute truth in classical physics had been dealt a significant blow, which was further compounded by Einstein's theory of relativity. In fact, the writer George Gamow called the first three decades of the 20th century "Thirty Years that Shook Physics" [22]. By the time Planck received the Nobel Prize in 1918, Neils Bohr had developed his theory of the atom, with which we deal in the next chapter. But as Planck himself pointed out in his Nobel lecture [5],

> To be sure, the introduction of the quantum of action has not yet produced a genuine quantum theory.

These theories were still almost a decade away and would further radicalize the understanding of physics. Moreover, Planck's quantum of action had not clarified the understanding of light itself.

The Conundrum

Young had performed what is now called the two-slit experiment almost a century before Planck, as we noted above. In this experiment, light with a relatively narrow frequency spread (today one does it with a laser) is passed through two slits in a partition and allowed to impinge upon a screen located behind the partition. For a wave moving through the two slits, there will be a phase shift of the light from one slit relative to the other as one moves away from the center line between the slits. This causes interference between the two parts of the wave, and one will see alternating light and dark regions. This occurs just as one throws a rock into a pond and sees ripples propagating away from the impact point. These ripples are alternating regions of deeper and shallower water. With two rocks, the ripples interfere causing a pattern equivalent to that of the Young experiment. If light were composed of classical particles, then they would go through one slit or the other, but would not create the interference pattern. Rather, there would just be two bands of light, one behind each slit.

A few years after Young, Augustin-Jean Fresnel performed a series of experiments on the diffraction of light [23]. Diffraction was an effect that had been observed far earlier [24], by Francesco Grimaldi, who actually coined the name. Following this, Huygens had developed the first classical wave theory for light [25]. With diffraction, if one passes light around a thin wire, a colored set of bands will appear on a screen placed behind the wire. If one uses an absorbing material on one side of the wire, so that light only passes around the other side of the wire, these bands disappear. So, the color pattern arises from the interference between light passing on either side of the wire. In a sense, this is an experiment equivalent to the two-slit experiment. Here, the diffraction of light by the wire depends upon the wavelength or frequency, so different wavelengths produce different interference effects, and thus the colored pattern appears. Similarly, if one passes a narrow band of light frequencies through a single hole, a series of rings is formed on a screen behind the hole. These interference fringes are the diffraction rings that occur from the edges of the hole. Fresnel's studies on diffraction confirmed Young's experiments showing that light was composed of continuous waves. Hence, before Newton as well as after him, we are confronted with experimental evidence for the continuity of waves, with their detailed behavior expressed in terms of Maxwell's equations for electromagnetism.

Yet Planck had shown that light was not continuous, but was composed of packets of energy called corpuscles, or later called photons. The energy of each photon was determined by Planck's constant and the frequency of light. This corpuscular nature of light would be clearly evident in the intensity of light, which represented the number of photons arriving on a screen per second. Then, Einstein showed that the quantization of light explained the photoelectric effect. This was used by several scientists to explain other experimental observations. By 1918, it was clear that light was quantized.

So, which was it? Is light a continuous wave, or is it quantized into small packets of energy? Certainly, one could tell the difference with various experiments. If the diffraction and interference experiments were repeated, light demonstrated clear, continuous wave-like properties. If the photoelectric experiment, and other similar experiments, were repeated, light demonstrated clear particle-like properties. This created a conundrum for the scientist. It appeared that the nature of light depended upon what experiments were performed. Are we to believe that light knows what the experiment will be, and adjusts its properties accordingly? Not likely! However, in 1909, Einstein discussed the fact that light had to have this dual nature and derived a formula for the energy fluctuations which had two terms—one arriving from the assumed particle form and one from the wave nature of light [26]. His result was confirmed some years later by Pascual Jordan, who started with a quantized wave and then computed the fluctuations [27] (we will also meet Jordan later). These results have been viewed with skepticism by most of the community, as it occurred during the argumentative days at the beginning of quantum mechanics proper (of which we speak later). Nevertheless, Jordan's results have been discussed in a more positive light in recent years [28]. Suddenly, with these latter results, we no longer have just wave *or* particle, but perhaps also now wave *and* particle.

Yet, we are faced with the fact that the particle, or wave, nature of the light is determined by the manner in which we put the experiment together. If we look for particle properties with the experiment, we find them. On the other hand, if we prepare an experiment to study interference or diffraction, we find wave-like properties. This begins to sound like the philosophical view of science and reality put forth vigorously in the late 19th century by Ernst Mach.

The Rise of Positivism

Ernst Mach was born in a part of the Austrian empire that is now the Czech Republic. Educated at Charles University in Prague, he began his career at the university in Graz, and worked on sensory perception. In 1867, he returned to Prague as a professor of Physics at Charles University. It was during his time in Prague that he investigated hydrodynamics and flow moving faster than the speed of sound. It is from these efforts that we know him from the Mach number, the ratio of an object's velocity to the local velocity of sound. But he gradually became famous as well for his philosophical views, which were sometimes called phenomenalism—that it, the reality in science is what we can observe and sense ourselves. This view precludes the presence of atoms and molecules because they were not observable. In fact, in 1895, Mach moved to the University of Vienna, where he took the position as professor of philosophy. By this time, his philosophy was more fully formed, and has often been referred to as *positivism*. The actual meaning usually depends upon which philosopher is giving us the meaning. Steven Brush has given this view [29]:

The word "positivism" has been used in various senses; it can denote simply the attitude that the methods of experimental science should be applied ..., instead of the methods of ... metaphysics. In this sense it merges with "naturalism," "realism," and 'materialism'.... But in philosophy, and to some extent in science itself, positivism is opposed to materialism, and comes to be synonymous with 'idealism" or "empiricism".

... the doctrine of ... Mach ... asserted that reality is a combination of sensations standing in a definite relation to each other.

And these views were shared by others, such that a significant part of the community held to the view [29]:

The most "advanced" and "sophisticated" theories were those that took a purely phenomenological viewpoint; scientific theories should deal only with the relations of observable quantities and should strive for economy of thought, rather than trying to explain phenomena in terms of unobservable entities by "flexible conjectures".

Thus, Mach was able to fully express his disdain and dislike of the work of Boltzmann and his statistical physics, particularly the kinetic theory of gases. Once, after a lecture by Boltzmann, Mach is said to have remarked that he did not believe that atoms existed, a statement which served to end all discussion of the talk and to cast disbelief in the minds of other attendees. What Mach meant by the remark is described by Holton as [30] "what Ernst Mach was attacking when he objected to the notion of atoms ... was not the phenomenic hypothesis of the atom ... [but] the concept of fundamental submicroscopic discreteness as against continuity." By this time, Mach had risen to serve as one of the leaders of the Vienna circle of philosophers preaching logical positivism. This latter group, and an equivalent group in Berlin, tended to dominate views of philosophy in central Europe. As a result, by the time of Einstein's photoelectric paper, it is believed that most physicists still rejected the reality of atoms. In fact, [31] "at the turn of the century good theoreticians were not yet familiar with statistical mechanics, a branch of theoretical physics that was indeed less than twenty years old." Support for Mach's philosophical view continued to grow, perhaps reaching its peak a quarter of a decade later.

Mach's view of physics made the experimental observation of an effect an important aspect of the real world of physics. His view can be poorly summarized by three statements: (i) physics should be based upon phenomena that can be directly observable; (ii) there is no absolute system of space and time, the only motion that one can observe is in reality relative motion; and (iii) if we see something that seems to be describable in an absolute system of space and time, it is instead the result of interactions with the large scale distribution of all matter in the universe. Even Einstein called this last statement Mach's principle, and said that relativity theory rests upon this principle, although he would eventually have to break away from such a philosophy.

Planck disagreed with Mach's views, and criticized his reluctance to accept the reality of atoms. As discussed above, Planck worked with thermodynamics in developing the black-body radiation formula and was also a strong proponent of Boltzmann. Indeed, one of the more obvious unobservables is a gas. We normally can't see the gas, but we know it is there from measuring other properties of the gas, such as temperature, pressure, volume, and density. These tend to be thermodynamic variables; hence studies of this field lead a person to accept the reality of atoms and molecules, no matter how small they may be. Einstein, who also worked early on thermodynamics, discussed in one of his 1905 papers the fluctuations in the density of atoms in a gas. These fluctuations produced measurable quantities from which one could sense the existence of the atoms. This was perhaps the beginning of his departure from positivism, although the ultimate break came with his development of general relativity. Here, Einstein had to believe that reality lay in the theory itself, even before any experimental evidence confirmed its validity. Holton has said that "it would not be possible to understand fully the philosophical progression of Einstein away from this early devotion to empiricism—unless one understood and made provision for the influence of such colleagues as Max Planck" [32].

Nevertheless, this schism in science in general, and in physics in particular, has lasted for quite some time. As late as the second half of the twentieth century, the argument persisted, with statements such as "Intuitive apprehension of the inner workings of nature, fascinating indeed, tend to be unfruitful. Whether they actually contain a germ of truth can only be found out by empirical verification; imagination, which constitutes an indispensable element of science, can never even so be viewed without suspicion" [33]. Another viewpoint of the period was "scientific theories should deal only with the relations of observable quantities" [29]. Fortunately, the rise of theoretical, and computational, science in the second half of the century has changed the view of most physicists, with the likely exception of those in quantum mechanics. Perhaps never has this change been so obvious than with the final discovery of the predicted Higgs' boson in late 2015 and the observation of Einstein's gravitational waves in 2016. These are especially significant for the large sums of money invested to try to find these theoretically predicted phenomena. But they also signal the reality of intellectually predicted quantities without the need for observation.

The Peace Maker

Through the story will run another individual, Paul Ehrenfest. Ehrenfest was an Austrian, from a lower-class family in Vienna. He began his university work at the Technical University in Vienna, but took most classes at the University of Vienna, where Ludwig Boltzmann was his advisor. As usual with the Germanic universities, students would take courses at various places in order to gain a wider understanding of the subjects in which they were interested. In 1901, Ehrenfest went to Göttingen, an important center for mathematics. There he met his future wife, Tatyana, who was a Russian student, also in mathematics. He finished his doctorate in 1904 and then married, and then returned to Göttingen. The untimely death of Boltzmann, by his own hand, both shocked him and opened an important door for him. The mathematician Felix Klein was the editor of an important encyclopedia on natural sciences, and had been counting on Boltzmann to write the definitive article on statistical physics. With the latter's death, he realized he had a strong adherent of Boltzmann on his doorstep and arranged for Ehrenfest to write the article.

Ehrenfest, over the years, had developed a health problem similar to that of Boltzmann, which would lead to bouts of depression. In his case, Ehrenfest would continually doubt his own ability to contribute to science, but he had an unusually strong ability to evaluate the work of others and to make it clearly understandable to others. His article on statistical physics would take a great deal of time (five years), but it would become the truly definitive work on the subject. While this was going on, though, he was not successful in finding a good university position, so that he moved to St. Petersburg for his wife. In 1911, the article was published, and the next year, he embarked upon another tour of the German universities, both to speak on statistical physics and to seek a position. This allowed him to meet Planck, Einstein, and Sommerfeld. The article had an effect, however. When Hendrik Lorentz retired in Leiden, and upon his recommendation, the university called upon Ehrenfest to fill his position. (Hendrik Lorentz was a Dutch physicist who shared the Nobel Prize with Pieter Zeeman in 1902 for the discovery and theoretical explanation of the Zeeman effect. Lorentz became professor of theoretical physics in Leiden when only 24 and made his name in electrodynamics and, later, in relativity theory. We will meet the Lorentz and the Zeeman effect often in the following chapters.)

In Leiden, Ehrenfest was a breath of fresh air, in that he was anything but the staid conservative professor, which was the norm. Rather, he took the position enthusiastically, creating a reading room where physics students could gather and talk and where the latest publications were available for them to read. He also created a social gathering for the students, often in his own home, to build camaraderie among them. In 1916–17, he made a major contribution to quantum theory with a major review in which he introduced the adiabatic approach [35, 36]. In this work, he proposed that quantum variables and their motion must eventually correspond to the equivalent classical variables and their motion.

In cases where Ehrenfest felt that students needed further information, he would encourage them to spend time at other universities with major professors. His growing acquaintance with the principle scientists of the time aided in this effort. Thus, for example, one of his students would go to Copenhagen to work with Niels Bohr.

Ehrenfest became quite popular in Leiden. He would hold major seminar series, inviting leading scientists to come and speak in Leiden. Thus, he would bring Niels Bohr to Leiden in 1919, and they too became friends. He arranged for Einstein to be appointed a visiting professor in Leiden in 1920. One important aspect of Ehrenfest, in later years, would be his continuing efforts to bring these two together in efforts to alleviate the growing scientific enmity between them. We will deal with this further in later chapters.

Modern Investigations

So, we measure the wave- or particle-like properties of light depending upon the experiment which we perform. Mach would have us believe that, in fact, the experiment that we chose made those properties real. They didn't exist before we did the experiment. Indeed, he likely would even have been repelled by the concept of light as a particle, and a very small particle at that. However, is this the proper way to think about light? Is it really to be interpreted based only upon the measurement we chose to make? Is there no reality prior to this measurement? As we will see, later founders of some of quantum theory would have us believe that this latter interpretation is the proper choice for quantum mechanics. Should this really be the case, or are they invoking Machian philosophy as a crutch to hide something? And what would that be? If we are to answer such a question, we have to look a little further into this wave–particle nature, and go back to our previous question. Is it wave *or* particle, or is it perhaps wave *and* particle. What do experiments of the intervening century of investigation tell us?

Already in 1909, Taylor [37], following an idea of J. J. Thomson [38], initiated experiments using very weak light. Thomson was studying the ionization of gases under ultraviolet and X-ray illumination, and he noticed that only a small fraction of the gas molecules were ionizing. This lead him to conjecture that the energy of the light was localized in Planck's packets, which also were *physically small in dimension*, so that they missed the majority of the molecules. This led him to suggest that light striking a metal surface would actually cause a "series of bright specks on a dark ground." He supposed that if the plate was moved further from the source, "we shall diminish the number of these [specks] ... but not the energy in the individual units." This went beyond Planck's quantization by also suggesting that the photons were physically quite small. Thompson estimated that, with illumination of 1.9 eV photons with an intensity of 10^{-11} W/cm^2, he would get about 3.3×10^7 photons per cm^2 per second, for which there would be about 1 photon for each 1000 cm^3. He finally concluded that "the structure of light would be exceedingly coarse." By attenuating the light in his measurements, Taylor could get a situation in which the photons came through one at a time and were separated by a large amount. Suppose that you want the individual photons to be 0.1 m apart. Since they are traveling at the speed of light, 3×10^8 m/s, this means that a photon arrives every 0.33 nanosecond. For visible photons at ~2.5 eV (Fig. 1.1), this means that the light beam must have a power of less than 1.2 nW. As we will see, this is certainly an achievable level of power, and the 0.1 m spacing of the photons means that there is never more than a single photon passing through the two slits. Taylor used a flame passing through a single slit as his source, and then diffracted the light around a needle. He attenuated the light with various smoked glass windows, which he calibrated first to determine the attenuation of each window. With the weakest light, the photographic plate had to be exposed for 2000 hours. He estimated that this corresponded to imaging a standard candle from 1 mile away with a 10 s exposure of the plate. Using the known properties for a standard candle, he calculated that, in the experiment, he had a minimum source of about 5×10^{-13} W/cm^2 impinging upon the plate, or that 1 cm^3 of air in front of the plate contained about 1.6×10^{-23} J of energy at any time. If the photons were 2 eV photons, this would mean that there was only a single photon for every 1/50 of a cubic meter. This is a lower light level, already achieved in 1909, than that discussed above. Taylor states that he saw no diminution in the diffraction pattern for any of his exposures, thus confirming Thomson's ideas. Even though light arrived at each photographic plate as a single photon at a time, the net pattern formed over time matched the expected wave interference patterns. Consequently, the light in this experiment was acting as both particle (upon striking the plate) and wave (when passing the needle).

From this and from later experiments on two slits we know that, what is observed is that each photon creates a single point on the screen behind the two slits. That is, a single photon impacting the screen does not create the interference pattern. But, after thousands of photons impact the screen, each one going through the two slits as an individual photon, the interference pattern emerges. (An excellent modern experiment demonstrating this appears on the Web [39].) This means that, while each photon creates only a single spot on the screen, it's passage through the two slits was such that its wave-like properties sensed both of the slits, so that the pattern of dots fully demonstrates the interference pattern. One could say that the interference, or wave nature, is a result of an ensemble of photons. One could conclude from this that the photon had both wave-like and particle-like properties. The measurement of the photon through its spot on the screen should invoke particle properties, in the Machian precept. But, it required its wave-like properties while traversing the two slits in order to have the ensemble lead to the interference pattern. One might even conclude that the photons had a reality of their wave and particle nature prior to the impact on the screen in order to explain this experiment. A noteworthy parallel to this experiment is that one might decide to look to see which slit the individual photon went through. It turns out that, when detectors are placed near the slits for this purpose, the photons devolve into pure particles, and the interference pattern disappears. This suggests that you cannot see both the wave properties and the particle properties simultaneously in an experiment. In fact, Niels Bohr, who we will meet in the next chapter, would express this later: "This point is of great logical consequence, since it is only the circumstance that we are presented with a choice of *either* tracing the path of a particle *or* observing interference effects" [40]. Thus, it seems to be a conundrum, and we can only measure one property or the other. In later years, though, there were suggestions for methods for measuring both the particle and the wave properties simultaneously [41, 42]. But sometimes the path from suggestion to observation can be quite long. As an aside, if we examine just the visual observation of the result, the human eye needs both wave and particle. The lens of the eye uses the wave properties to focus the light. The rod, cone, and photosensitive ganglion photoreceptor cells react to the particle nature of light. This action occurs without us knowing it or consciously controlling it.[b]

A modern approach to this would use coherent single photon sources to maximize the effect. Single-photon states apparently were first created in 1986 [43, 44]. It was not until just a few years ago that a series of measurements used single photons and the properties of quantum optics, along with the concept of "weak" measurements (to be discussed later). These "weak" measurements did not destroy the quantum properties of the photons, so that the experimenters apparently could observe both the specific paths and the interference pattern [45–47]. Interestingly enough, Kocsis et al. [45] were able to reconstruct the trajectories of the photons, and these trajectories look surprisingly like those expected from a modern interpretation which asserts the particle *and* wave theory (which we will discuss in the next chapter). This result has recently been repeated [48]. Finally, the wave–particle duality of light has actually been imaged [49]. These latter authors created a plasmonic confinement by vaporizing a nanowire, and then used electrons to image the array of individual photons. Thus, we are led to believe, from the recent advances in measurement capability, that one can observe both the particle and the wave nature of light simultaneously.

In the light of modern measurements, it would appear that Bohr's comments above were, at best, premature. As Pais has said, "Bohr was trying to allow for the simultaneous existence of both particle and wave concepts, holding that, though the two were mutually exclusive, both together were needed for a complete description" [50]. Nevertheless, it appears such a view was not altogether correct. Rosenfeld summarized the situation as follows [51]:

---

[b]Thanks to Alex Kirk for calling this to my attention.

Now, we know that the duality of behavior of light is irreducible: its paradoxical character, from the classical point of view, has forced us to a radical revision of the most fundamental conceptions of physics, and brought about a revolution in our view of the universe comparable to that which Galileo and Newton initiated when they created modern science.

The situation today, while a little clearer due to the recent experiments, is probably still debatable, and I doubt that the paradoxical nature has yet been fully resolved. It would seem, however, a view in which the experiment determines which property of light we observe can no longer be defended, especially if both properties can be simultaneously measured. If Machian logic finally fails for this aspect of the quantum revolution, it is very likely that it will eventually fail for all aspects of the quantum revolution. But, this jumps to a judgement when we are still early in the development of this revolution.

References

1. Protokollbuch19001214Planck-1.pdf, https://www.dpg-physik.de/veroeffentlichung/archiv/index.html.

2. M. Planck, *Scientific Autobiography and Other Papers*, trans. F. Gaynor (William and Norgate, London, 1950), p. 7, cited in A. Pais, *Neils Bohr's Times, in Physics, Philosophy, and Polity* (Clarendon Press, Oxford, 1991), p. 80.

3. T. Young, *Course of Lectures on Natural Philosophy and Mechanical Arts*, Vol. 1 (J. Johnson, London, 1807), Lecture XXXIX.

4. M. Planck, *Ann. Phys.* **4**, 553 (1901); trans. K. Ando, and available at http://dbhs.wvusd.k12.ca.us/webdocs/Chem-History/Planck-1901/Planck-1901.html.

5. M. Planck, Nobel lecture, http://www.nobelprize.org/nobel_prizes/physics/laureates/1918/planck-lecture.html.

6. W. Wien, *Philos. Mag.*, Ser. **5**, 43, 214 (1897); trans. *Ann. Phys.* **58**, 662 (1896).

7. J. W. S. Rayleigh, *Philos. Mag.*, Ser. 5, **49**, 539 (1900).

8. J. H. Jeans, *Philos. Mag.*, Ser. 6, **10**, 91 (1905).

9. H. Beckmann, Inaugural Dissertation, Tübingen (1898), cited at http://www.chemteam.info/Chem-History/Planck-1901/Planck-1901.html.

10. O. Lummer and E. Pringsheim, *Verh. Dtsch. Phys. Ges.* **12**, 215 (1899).

11. O. Lummer and E. Pringsheim, *Ann. Phys.* **3**, 159 (1900).

12. H. Rubens and F. Kurlbaum, Sitzungber. *Imp. Preuss. Acad. Sci.*, pt. 2, 929 (July–December 1900).

13. H. Rubens and F. Kurlbaum, *Ann. Phys.* **4**, 649 (1901).

14. H. Hertz, *Ann. Phys.* **267**, 983 (1887).

15. A. Stoletow, *Philos. Mag.*, Ser. 5, **26**, 317 (1888); trans. of *Comp. Rend.* **CVI**, 1593 (1888).

16. P. Lenard, *Ann. Phys.* **313**, 149 (1902).

17. A. Einstein, *Ann. Phys.* **322**, 132 (1905).

18. J. Franck and G. Hertz, *Verh. Dtsch. Phys. Ges.* **16**, 457 (1914).

19. W. Duane and F. L. Hunt, *Phys. Rev.* **6**, 166 (1915).

20. D. L. Webster, *Proc. Natl. Acad. Sci.* **2**, 90 (1916).

21. Von E. Warburg, *Sitzungber. K. Preuss. Akad. Wiss.* (Jan.–Jun. 1912), pp. 216–225.

22. G. Gamow, *Thirty Years that Shook Physics* (Double Day, Garden City, NJ, 1966).

23. A.-J. Fresnel, *Ann. Phys. Chem.*, Ser. 2, **1**, 239 (1816).

24. F. M. Grimaldi, *Physico mathesis de lumine, coloribus, et iride, aliisque annexis libri duo* (Bonatti, Bologna, 1665), pp. 1–11; trans. G. Bernia (Dawson's of Pall Mall, London, 1966). The Latin original is online at http://193.206.220.110/Teca/Viewer?an=300682.

25. C. Huygens, *Traite de la Lumiere: ou sont où sont expliquées les causes de ce qui luy arrive dans la Reflexion, and dans la Refraction* (Peter van der Aa, Leiden, 1690).

26. A. Einstein, *Phys. Z.* **10**, 817 (1909). An English translation is available at https://en.wikisource.org/wiki/The_Development_of_Our_Views_on_the_Composition_and_Essence_of_Radiation

27. M. Born, W. Heisenberg, and P. Jordan, *Z. Phys.* **35**, 557 (1926). An English translation is in B. L. van der Waerden, *Sources of Quantum Mechanics* (Dover, Mineola, NY, 1967), pp. 321–385.

28. See, e.g., A. Duncan and M. Janssen, *Stud. Hist. Phil. Mod. Phys.* **39**, 634 (2008).

29. S. Brush, *Graduate J.* **7**, 477 (1967).

30. G. Holton, *Thematic Origins of Scientific Thought* (Harvard Univ. Press, Cambridge, 1973), p. 100.

31. A. Pais, *Neils Bohr's Times, in Physics, Philosophy, and Polity* (Clarendon Press, Oxford, 1991), p. 79.

32. G. Holton, *Thematic Origins of Scientific Thought* (Harvard Univ. Press, Cambridge, 1973), p. 39.

33. E. J. Dijksterhuis, *The Mechanization of the World Picture*, trans. C. Dijksterhuis (Clarendon Press, Oxford, 1961), p. 304.

34. S. Brush, *Graduate J.* **7**, 477 (1967).

35. P. Ehrenfest, *Ann. Phys.* **51**, 327 (1916).

36. P. Ehrenfest, *Philos. Mag.*, Ser. 6, **33**, 500 (1917).

37. G. I. Taylor, *Proc. Cambridge Philos. Soc.* **15**, 114 (1909).

38. J. J. Thompson, *Proc. Cambridge Philos. Soc.* **14**, 417 (1907).

39. http://www.tnw.tudelft.nl/en/about-faculty/departments/imaging-physics/research/researchgroups/optics-research-group/education/experimental-projects/photons-in-an-optical-interference-experiment/ .

40. N. Bohr, in *Albert Einstein: Philosopher-Scientist*, ed. by P. A. Schilpp (The Library of Living Philosophers, Evanston, IL, 1949), reprinted in J. A. Wheeler and W. H. Zurek, *Quantum Theory and Measurement* (Princeton Univ. Press, Princeton, 1983), p. 9.

41. W. K. Wooters and W. H. Zurek, *Phys. Rev. D* **19**, 473 (1979).

42. L. S. Bartell, *Phys. Rev. D* **21**, 1698 (1980).

43. C. K. Hong and L. Mandel, *Phys. Rev. Lett.* **56**, 58 (1986).

44. P. Grangier, G. Roger, and A. Aspect, *Europhys. Lett.* **1**, 173 (1986).

45. S. Kocsis et al., *Science* **332**, 1170 (2011).

46. R. Menzel et al., *Proc. Natl. Acad. Sci.* **109**, 9314 (2012).

47. P. Kolenderski et al., *Sci. Rep.* **4**, 4685 (2014).

48. D. H. Mahler et al., *Sci. Adv.* **2**, 1501466 (2016).

49. L. Piazza et al., *Nat. Commun.* **6**, 6407 (2015).

50. A. Pais, *Neils Bohr's Times, in Physics, Philosophy, and Polity* (Clarendon Press, Oxford, 1991), p. 309.

51. L. Rosenfeld, in *Physics in the Sixties*, ed. by S. K. Runcorn (Oliver and Boyd, 1963), pp. 1–22, reprinted in *Selected Papers of Leon Rosenfeld*, ed. R. S. Cohen and J. J. Stachel (Reidel, Dordrecht, 1979).

Chapter 2

The Arrival of Bohr's Atomic Theory

The belief that small things like atoms existed is as old as the ancient Greeks and the ancient Indians (here I refer to the people of the subcontinent in Asia). Of course, they could not see these things and did not have the capability to try to find them experimentally. But at least one of the philosophies extant at that time had atoms as the smallest unit of matter. We have to remember that scientific proof at that time followed mostly from consistent thought which could withstand challenges from other great thinkers. Hence, the idea that there was such a smallest unit carried forward over the centuries as a philosophical idea. A firmer idea of the reality of the atom came in the early 1800s from John Dalton, who studied the various constituents of compound gases and materials [1]. For example, he studied how nitrogen gas and oxygen gas could combine, and found two possibilities. He found that it could combine as one part of oxygen along with one part of nitrogen (NO, nitric oxide), or with two parts of nitrogen (N_2O, nitrous oxide). From the weight of the various gases, he determined that the fractional weight of nitrogen in the two compounds differed by a factor of 2. From this, and studies of other molecules, he concluded that each atom of nitrogen had a very specific weight—the atomic weight. Hence, he put reality into the idea of the atom by giving it a weight. In fact, he later published

The Copenhagen Conspiracy
David Ferry
Copyright © 2019 Pan Stanford Publishing Pte. Ltd.
ISBN 978-981-4774-75-8 (Hardcover), 978-1-351-20723-2 (eBook)
www.panstanford.com

a table of elements arranged according to their atomic weights, but it would take more than a century to confirm his views of the atom.

In 1827, Robert Brown, a botanist, studied small pollen particles suspended in liquid and noticed their random motion [2], which subsequently became known as Brownian motion. In Einstein's second 1905 paper, he provided a statistical theory for this motion [3]. This work inspired the French physicist Jean Perrin to undertake a study of the mass and dimensions of individual atoms [4], which verified Einstein's prediction, and also finally verified Dalton's theories of the atom. It was clear that atoms were very real objects, with definitive masses and sizes based on the particular element, despite the views by Mach in the previous century.

The ancient philosophical view that atoms existed was thus the correct view. By early in the twentieth century, their existence and properties had been confirmed and measured. Importantly, a theoretical development based upon the statistical physics of Boltzmann had been a key part of this confirmation. Mach would have turned over in his grave. The question now turns to the point that if atoms are real, what is their structure? They cannot just be blobs of matter. Scientists needed to know more, and this information was beginning to arrive already at the end of the nineteenth century. It had been known for quite some time that one could create two types of electrical charge, positive and negative. Earlier work had suggested that large atoms were made up of smaller units, but it was believed that the smallest of these units would be about the size of the hydrogen atom (and probably the hydrogen atom itself). We mentioned Lenard in the previous chapter, where he had used a cathode ray tube to measure the velocity of the emission from the cathode. This tube was a system in which a potential could be placed between a pair of plates, and emission could be seen from the negative electrode. Although called Lenard rays at the time, they came to be called cathode rays, and most physicists believed they were composed of some kind of liquid. Adding light increased the emission, so the results of the photoelectric effect suggested that the rays were actually particles. By evacuating the tube, J. J. Thomson discovered that the particles could be deflected in electric and magnetic fields, and thus he could discover both their mass from a magnetic field and their charge to mass ratio from crossed electric and magnetic fields [5, 6]. He concluded that the mass of this particle, to be called the electron, was more than 1000 times smaller than the mass of the hydrogen atom (and we would learn later that it was actually 1837 times smaller than the proton, one of the constituents of the nucleus). Consequently, he had discovered the first subatomic particle, which meant that the atom could be subdivided. Not only were atoms real, but they now had to be considered to be constituted of much smaller objects. The world had suddenly become much more complicated. Not only were there subatomic particles, but the mass and charge of the electron had been measured.

Earlier in his career, Thomson had studied the vortex theory of the atom proposed by another Thomson—William Thomson, Lord Kelvin. This theory proposed that the atom was a vortex—a spiraling singularity like a whirlpool—in the ether. From many mathematical studies of the theory came modern topology and knot theory. But with the ether no longer necessary after Einstein's relativity theory, this approach to the atom went away as well. So, J. J. Thomson turned his vision elsewhere and proposed that the atom was composed of negatively charged electrons dispersed throughout a solid volume of positive charge, which is usually referred to as the nucleus [7]. This model is often thought of as the *plum pudding model*, or perhaps it is more familiar if we think of it as a blueberry muffin, where the dispersed blueberries play the role of the electrons. While we no longer take this as the proper model for the atom, it did solve some critical issues in classical electromagnetics. If the electrons were outside the positively charged nucleus, then there would be a large attractive force pulling the electrons back to the nucleus. Hence, the structure would collapse to the plum pudding model. This issue would become critical as the ideas on the structure of the atom progressed.

The next development would come from Ernest Rutherford, already a Noble laureate. Earlier, Geiger and Marsden had studied the scattering of α particles from heavy metals, and found that most were scattered only by very small angles, but a few were scattered through quite large, and unexpected, angles [8, 9]. They further showed that the large scattering angle was dependent on the atomic weight of the metal. Rutherford then calculated the scattering of a charged α particle from a charged point and deduced that the large angle scattering could only occur if most of the mass and (positive) charge were localized [10]:

> In comparing the theory outlined in this paper with the experimental results, it has been supposed that the atom consists of a central charge supposed concentrated at a point, and that the large single deflections . . . are mainly due to their passage through the strong central field.

In further considerations, he concluded that the charge in the central point had to be highly concentrated, and the main interaction was due to the Coulomb force between the central nucleus and the charged α particle. His conclusions were independent of the sign of the central charge and, indeed, he was not sure whether it was the electrons or the positive charge in the central point. But since the amount of charge was proportional to the atomic weight, he concluded that it was most likely that the central point, the nucleus, was the positive charge. Moreover, a positively charged nucleus would explain the high velocity of small masses emitted from atoms under radiation. A further complication, though, is that this central point had to be much smaller than the assumed size of the atom. If this were the case, then it was likely that most of what was thought of as the volume of the atom was, in fact, empty with only the collection of very light electrons filling this space. But wouldn't the electrons be drawn into the nucleus by the strong Coulomb interactions between them and the positive charge in the nucleus? Rutherford noted that Nagaoka [11] had modeled a planar atom in which the electrons all resided on a circle around a central positive nucleus. The latter had shown that this was stable provided the attractive force between the electrons and the nucleus was large enough, and the electrons all had about the same velocity. Rutherford noted that the large angle scattering would be the same whether the atom was a disk or a sphere, but he thought it more likely that the electrons would be in orbits much like our planetary system. In fact, Nagaoka had considered already that the electrons may lie in many rings, all of which may or may not lie in the same plane, so it could be a planetary-like model or a more spherical model. He noted that the many rings would align better with studies of the optical absorption measurement in atoms, which showed a series of very sharp lines in the spectra, which he connected to the rings of his atomic model.

Absorption and Emission

The absorption of light in atoms, as mentioned above, is not a continuous process, but consists of a series of distinct lines. If an atom is ionized, then the emission of light by the atom as the electrons recombine is also a set of distinct lines, with the emission and absorption lines coinciding. This process is a modern method of characterizing what atoms exist in the atmosphere of a planet or star. One of the early pioneers was the Swedish physicist Anders Jonas Ångström. In the middle of the nineteenth century, he developed tools with which he studied the emission spectrum of the sun, and determined that hydrogen, as well as other atoms, were in this spectrum [12], building upon the work of, e.g., Fraunhofer. In 1868, he published what is known as the great map of the solar spectrum [13]. Working with the data from these publications, Johann Balmer devised a formula to empirically fit this data [14]. Balmer was a Swiss mathematician, and this seems to be his greatest contribution, done when he was already in his 60s. He wrote the formula in terms of the wavelength of light, λ, discussed in the last chapter, as

$$\lambda = \frac{Bn^2}{n^2 - 2^2}, \tag{2.1}$$

where B is a constant and $n = 3, 4, 5, \ldots$. Of course, he was not clear why the second number in the denominator was 2, but it fit all the data in the visible spectrum to which he had access.

Working almost independently, Johannes Rydberg, a Swedish physicist, developed a somewhat more general formula, as he was trying to fit to a wide range of materials. He recognized that Balmer had found a special case of a more general formula that he had created [15]. Rydberg's formula read

$$\frac{1}{\lambda} = RZ^2 \left(\frac{1}{m^2} - \frac{1}{n^2} \right), \tag{2.2}$$

which reproduced Balmer's formula when $m = 2$. Here, R is known as the Rydberg constant, which he thought was the same for all materials, and Z is the atomic number (the number of protons in the atom). For the case of hydrogen, where $Z = 1$, the two equations are connected by $B = 4/R$.

As Balmer's series was mainly in the visible, it is natural to understand that it was this part of the hydrogen spectrum that was the first one found and remained the only known set of emission lines for a great many years. Finally, after the turn of the century, the German physicist L. C. H. Frederick Paschen found a series in the infrared that corresponds to $m = 3$ [16]. A few years later, the American physicist Theodore Lyman found the series in the ultraviolet corresponding to $m = 1$ [17]. So, it is clear that the Rydberg formula above is quite general in its application and very precise for hydrogen. Yet, it is based upon fitting to experimental data. In the early years, there was no theoretical basis for this formula, although Nagaoka had created an opening toward one [11]. Nagaoka knew about Balmer's and Rydberg's formulae, so it is surprising that he did not try to reproduce one of these formulae. However, he never introduced Planck's constant in any way, so perhaps it is not so surprising.

The problems in physics had not gone away even with this body of relatively new physics. Indeed, it had only become exacerbated. Perhaps it is best expressed by Leon Rosenfeld [18]:

> Beginning in 1909, a deep split appeared between the radical point of view of Einstein and the moderate tendency of Planck. The contentious question raised by Einstein's [1905] ... work was indeed one of the most fundamental: this problem of the structure of light was to continue to dominate the entire later development of the quantum theory ... the major problem was to determine the relationship between the corpuscular and the wave properties of radiation ... But if the concept of light quanta as spatial localizations of energy was insufficient to provide the solution, it is not because the idea was too radical, but rather because it was not radical enough.

Planck and, to some extent, even Einstein had not yet made the full evolution beyond classical physics, and the purely classical atomic views of Rutherford and Nagaoka did not provide the explanation of the absorption and emission spectra of hydrogen (much less any heavier atom). To this world of confusion and contention, the next big step came from a young Dane.

Niels Bohr

Bohr was born into an affluent family in Copenhagen, as his father was a professor of physiology at the university. The family was quite well situated, both politically and socially as [19] "the Bohr family belonged to the Danish elite at the time of Niels' birth." In fact, before Niels' father became a professor (of physiology), the family mansion was across the street from the royal palace in Copenhagen [20]. With such political backing, his career was certain to blossom in Copenhagen. Notably, his younger brother Harald became professor of mathematics at the university and was obviously the mathematician in the family. The problem for both boys is that there was no infrastructure in physics and mathematics in Denmark, and only one professor of physics at the university. Nevertheless, Niels was educated locally, receiving both his graduate and undergraduate degrees from Copenhagen University. His doctoral thesis dealt with the free electron theory for metals, in which the electrons are not bound to the atoms, but move throughout the material much like a gas. This work built upon prior work by Drude and Lorentz, but he was unable to achieve an explanation of the experimental Hall effect, in which a magnetic field applied normal to the current flow direction led to a transverse electric field normal to both the current and the magnetic field. Perhaps this failure is a direct result of the lack of a physics infrastructure at the university, because the consideration of the free electron theory without a finite size sample cannot give the results necessary for the Hall effect. What is needed is just this finite sample so that the Lorentz force on the electrons (due to the fact that the current and magnetic field are perpendicular to one another) can build up non-equilibrium charge between the two opposite sides of the sample. It is this charge inhomogeneity that creates the voltage between the two sides, which is known as the Hall voltage and is the primary observable in the Hall effect. So long as one assumes that the charge must remain homogeneous, a great many effects will not be observed. For example, the transistor action so necessary in modern electronics depends primarily upon the control of charge inhomogeneity.

In his university courses, he attended the required lectures on philosophy by Harold Høffding, a noted philosopher. In addition, he was a long-time family friend and often visited the Bohrs throughout Niels's younger years. It is generally believed that Høffding was strongly influenced by Kierkegaard and finally settled into positivism. As Holton put it [21], "Kierkegaard's stress on discontinuity between incompatibles, on the 'leap' rather than the gradual transition" would prove to be influential in Bohr's later work. But Høffding's move to positivism also meant that Mach's general philosophy would be imprinted on the young Bohr, as Høffding held that the basic philosophical questions could be boiled down to the nature of consciousness. This led him to believe that the nature of reality was unknowable and we could only get glimpses of it through our conscious observations, that is through experiment. One could not get much closer to Mach than this, and much further from the approach of the ancient Greek philosophers. We will see this philosophical viewpoint appear again and again in Bohr's statements and work. Nevertheless, it has been said that Bohr has denied any influence from Høffding on his own philosophy [22]. Perhaps this is because Bohr was interested in establishing his own philosophical theories based on his ideas about quantum theory, and did not want anyone believing it was based on the work of others.

Following Bohr's doctoral work, he received a fellowship from the Carlsberg foundation[a] which would allow him to travel to England for further research. Initially, he went to the Cavendish Laboratory at Cambridge to work with J. J. Thompson, but this did not seem to be a good match. Subsequently, at the invitation of Rutherford, he moved to Victoria University in Manchester. There he studied Rutherford's planetary model of the atom, but was disturbed

---

[a]This is the famous brewery in Copenhagen. Bohr would ultimately also move to the Carlsberg mansion provided by the brewery to an eminent person in Copenhagen.

by the question of why the electrons didn't simply collapse into the nucleus as they radiated their energy (by the aforementioned bremsstrahlung). This led him to ask whether the electrons would be restricted into very well defined energy levels using the same idea of quantization that Planck had used. He was still working on the problem when his fellowship expired and he returned to Denmark, and perhaps more importantly, got married. He had a position at the university, but left it to work on his atomic model (and, in addition, postponed his honeymoon until his paper was finished) [20]. The first paper was finally published in 1913 [23]. Surprisingly, Bohr had not learned about the Balmer and Rydberg formulae until about this time, but learned just in time to use the experimental results in his paper. This would be a fortunate addition, because it clarified his thoughts and provided the final justification for the paper. The paper itself was published in English in one of the British journals, and "communicated" to the journal by Rutherford on his behalf. This was certainly done in anticipation of the reception of the paper, which he viewed as a major breakthrough.

After publication of the paper, Bohr and his wife took their delayed honeymoon and traveled through Britain before taking up a new position in Copenhagen. Still, he quickly returned to Britain to again spend time with Rutherford, before returning, in 1916, to the Chair of Theoretical Physics at the University of Copenhagen, a new position which had been created specifically for him. This could not have occurred without both his scientific accomplishment and some political pressure. The former came from the fame his papers brought him, while the second arrived from the stature of his family and their connections. These further led to the establishment of the Institute of Physics in 1918, with Bohr as the director. This was a fairly important accomplishment for someone who was still only 33 years old. While this certainly added to the self-esteem of the young scientist, it also finally created an infrastructure for physics, both at the university and in Denmark in general. Indeed, it would be both Niels Bohr and the Institute of Physics that became a central focus for young scientists interested in the new atomic physics.

Bohr's Atom

The aim of Bohr's "great" paper was to try to explain how Rutherford's planetary model of the atom could exist. The problem of the electron collapse due to energy radiation was foremost in his mind, and he recognized [23] "the inadequacy of the classical electromagnetics in describing the behaviour of systems of atomic size." The major point is that such classical models had no space for Planck's quantization of light. He recognized that the introduction of this quantization might change the dynamics of the electrons around the nucleus. His mathematical considerations are simple and quite straightforward, and I shall follow his lead here.[b] He began by recognizing that an electron of charge $-e$ would be attracted to a nucleus of charge Ze, and moving this electron from far, far away to a distance r in the attractive potential would result in the electron having energy

$$E = \frac{Ze^2}{4\pi\varepsilon r},\tag{2.3}$$

where E is the energy and ε is the permittivity of free space. Now, this energy of the electron will consist of its kinetic energy in orbiting around the nucleus, which can be written as

$$E = \frac{1}{2}mv^2 = \frac{1}{2}mr^2\omega^2,\tag{2.4}$$

where m is the mass of the electron and ω is the radian frequency ($= 2\pi f$) at which the electron moves around the nucleus. From these two equations, he could find the frequency and radius as a function of the energy:

$$\omega = \sqrt{\frac{2}{m}}\left(\frac{4\pi\varepsilon}{Ze^2}\right)E^{3/2}, \qquad r = \frac{Ze^2}{4\pi\varepsilon E}.\tag{2.5}$$

These last two equations are Bohr's starting point for his "theory." He then takes into account the energy radiation. The electron outside the atom has a potential energy relative to this energy level, and in being trapped by the nucleus, it is assumed that as it falls into the potential well, it radiates the energy of this fall by photons. As

---

[b]However, I will use modern notation and the MKS system of units rather than the archaic ones used at that time.

he says [23], "then, from Planck's theory, we might expect that the amount of energy emitted by the process considered is equal to nhf, where h is Planck's constant and n is an entire number." Here, an "entire number" is known today as an integer. Hence, he makes the assumption that

$$E = n\frac{h\omega}{2\pi}. \tag{2.6}$$

Using this equation with the equation for ω just above, one then finds the energy level of the electron as

$$E = \frac{m}{2}\left(\frac{Ze^2}{2h\varepsilon}\right)^2\frac{1}{n^2}, \tag{2.7}$$

which is the energy for the n-th electron level. This end result is correct, but as we will see, other parts of the model are not correct. The interesting thing, of course, is that the prefactor, all the terms on the right except for that in n^2, give the value of the Rydberg constant R. There is an important point here that I will discuss again later, and that is the fallacy of the assumption in Eq. 2.6. Bohr is going to assert that the observed spectra of the hydrogen atom is due to transitions between the energy levels given in Eq. 2.7 (each value of n gives a level). But this is not consistent with the assumption in Eq. 2.6, where he assumes that it takes n photons to get to the n-th level.

Now, the problem here is how can one say that the radian frequency of the photon is exactly equal to the radian frequency at which the electron orbits the nucleus. This is quite an assumption, and requires a great leap of faith. This idea is already nearly present in Nagaoka's work [11]. As mentioned, Rutherford cites this work at the point where he puts the planetary model together. It is likely that Bohr would have become familiar with this work while he studied the Rutherford model of the atom. But Bohr does not mention this work in his paper. In fact, Bohr is sloppy here, in that he actually assumes that the frequency of the photons is half that of the electron's frequency about the nucleus, and then messes up the conversion from frequency to radian frequency. The above discussion cleans this up, but one is then left with the requirement that the photon frequency is equal to that of the orbiting electron. Without this, one does not get the connection to the Rydberg formula.

Bohr then goes on to discuss the connection with optical spectra. The principal assumptions that he makes are [23]

(1) That the dynamical equilibrium of the systems in the stationary states can be discussed by help of the ordinary mechanics, while the passing of the system between different stationary states cannot be treated on that basis.

(2) That the latter process is followed by the emission of a *homogeneous* radiation, for which the relation between the frequency and the amount of energy radiated is the one given by Planck's theory.

The first assumption is to really justify the approach outlined above. The second implies that a photon will be emitted when the electron moves from one energy level to the other, and it is this difference in energy values that now gives us the Rydberg formula. He then notes that if the lowest energy level involved corresponds to $n = 2$, then you arrive at the Balmer series.

Bohr goes on in this paper to conclude that the angular momentum of each electron in the atom is given by a multiple of $h/2\pi$, so that not only the energy levels, but also the angular momentum is quantized. It turns out to be somewhat more complicated than this, and Bohr should have known that the angular momentum quite likely would be more complicated as the central force problem had been discussed by Newton and both Lagrange and Hamilton had discussed their approach to mechanics a century earlier. We may note that the value of the angular momentum comes from the above equations, and an attempt to force a more correct form would have upset his results and the connection to the Rydberg constant.

Bohr would publish two more papers that year connected with this problem, in which he discussed other atoms and molecules. But this first paper is the crucial one which defines his work. However, already in 1912 with a preliminary version of the paper, Rutherford had asked how the electron knew to which state it must transition. While successful in explaining the Rydberg formula, the paper itself is amazingly quiet about the transition periods where the electron moves from one energy level to another. Bohr never used the words "quantum jump" in the paper, but it became very much in vogue shortly after this. Because he was dealing only with the stationary states (the energy levels), the consideration of how an electron made the transition from one level to the next did not seem to be of consideration. If this transition led to the emission or absorption of a single quantum of light, then one might be led to believe it should be instantaneous. Otherwise, one had to ask about fractions of a photon, if one remained of the view that the photon was a particle and not a wave. This point would become of philosophical interest in later years, and the requirement of emitting a single photon would give rise to the so-called quantum jump of the electron from one level to another.

Here, though, we are facing the fact that this was not a solid theory. Rather, it was a hand-waving exercise which fortuitously could yield an understanding of the source of the Rydberg formula, and by connection also the Balmer formula. It explained the series of lines and fit the experiments that we have discussed so far. But, if you believed that reality existed without the need for observers, then Bohr's work was unsatisfying. So, now the dilemma arises. As Pais asked [24], "What did Bohr really do? How could it happen that two generations would accord Bohr's influence the highest praise while the next one hardly knew why he was a figure of such high significance?" He had not developed a theory, but had come up with a formula that fit experimental observations and gave a value for an important constant. Perhaps it is at this point that Bohr's philosophy became so important. For, as Holton has said [25], "in Bohr's sense, a 'phenomenon' is the description of that which is to be observed *and* of the apparatus used to make the observation." Whether or not Bohr felt he was influenced by Høffding and Mach, he was going to adopt their world. It was the reality of the measured atomic spectra that defined the reality of his work. Reality arrived with the measurement and different measurements could have different realities, as we shall see in future pages. If the measured atomic spectra defined reality, then his theory was good to go. There did not need to be any underlying theoretical derivation—his approach fit the experimental results and thus was as real as it gets. Mach would have smiled.

Almost at the time of the first paper, Bohr was having second thoughts about the use of Planck's relationship. Mach had railed against the idea of the atom as an incarnation of discontinuity in science. And here, Bohr himself had brought in Planck's discontinuity as a key element of his atom. In his second thoughts, which will continue for quite some time, he thought about getting rid of this discontinuity. Thus, in his second paper, he turns around his derivation [26]. Now, he begins with quantization of the angular momentum, with a form he found in the first paper almost as an afterthought. Now, he can use this to derive the formula for the energy levels. And, he can do it without mentioning Planck's quantization of the photon. From later events, we might think that he now feels he has returned to continuity, but this just doesn't follow from quantization of the angular momentum.

In the first paper, Bohr had additionally used the charge (Ze) on the nucleus as a representation of the attractive force. He received some confirmation the same year from work done by H. G. J. Moseley [27], an English scientist who showed that the atomic number in fact was the amount of positive charge on the nucleus. Moseley was working on the radiation of X-rays from atoms. Prior to this, the atomic number was just the place the atom was put in a periodic table. Moseley was able to show that the frequency of X-rays emitted (or absorbed) by an atom was proportional to the square of a number which was the amount of charge in the nucleus. This is now known as Moseley's Law. In the case of hydrogen, for which Bohr's calculations were primarily suited, this worked out just fine. Moseley noted [27]:

> We have here a proof that there is in the atom a fundamental quantity, which increases by regular steps as we pass from one element to the next. This quantity can only be the charge on the central positive nucleus ... [this quantity] is the same as the number of the place occupied by the element in the periodic system.

In a subsequent paper, he noted that van den Broek had put forward the view that the charge carried by the nucleus is always an integral multiple of that carried by the hydrogen atom [28]. Moseley then concluded [29]:

> There is every reason to suppose that the integer which controls the X-ray spectrum is the same as the number of electrical units in the nucleus, and these experiments therefore give the strongest possible support to the hypothesis of van den Broek.

There is speculation that Moseley would have received a Nobel Prize for this work had it not been for the fact that he was killed at Gallipoli.[c]

Still, there was trouble in paradise. Anyone examining the theory should have noticed the inconsistency in Bohr's first paper. This lay in Eq. 2.6, as mentioned above. Without this equation, Bohr could not have found the proper energy levels and the Rydberg formula. Yet, this equation is inconsistent with the results. It works well enough for the lowest energy level, $n = 1$. An electron coming from free space and falling into the potential well around the nucleus would emit a single photon with a frequency given by the lowest energy level. In fact, though, this isn't quite true. If we have to conserve energy, there is a problem. In the usual parlance, the energy level of free space is taken to be zero. Thus, the energy level of the lowest energy level, from Eq. 2.7 is just $-R$. However, once the electron is in the lowest energy level, it still has kinetic energy, so that it must have had precisely this kinetic energy in free space before emitting the photon and falling into the energy level. This is rather a strict requirement, which Bohr is rather silent about. Worse, Eq. 2.6 fails entirely for any other energy level. Our electron in free space that is trapped into the second energy level still emits only a single photon, not the two that are required by this equation. The optical spectra of the atom confirms this. Of course, we might conclude that the electron gets to the second energy level by a two-step process, first absorbing a photon to get to a higher level (above the second level), and then emitting a second photon to drop to the second level. But these latter two photons would not have the same frequency and

---

[c]Gallipoli was an attempt by the allies, beginning in 1916, to invade Turkey and seize control of the Dardanelles. It failed.

would not agree with the lines of the optical spectrum and also of Eq. 2.7. So, they wouldn't satisfy Eq. 2.6. We are left with the fact that the assumption of Eq. 2.6 is not consistent with the results or the experiments. Yet, without this equation, Bohr has no theory. It is absolutely necessary to make this assumption in order to arrive at Eq. 2.7 from his starting point.

Then, the energy levels didn't quite fit the actual frequencies that had been measured in the spectrum for hydrogen. Sommerfeld [30] pointed out that Bohr had assumed that the nucleus was stationary, and, if one did the mathematics properly, the electron mass should be replaced with the reduced mass for the joint motion of the electron and the nucleus. Then, one also needed to consider that perhaps the orbits of the electrons were elliptical, the results would better fit to the experimental measurements. Further, Bohr had the same problem with magnetic field that he had in his doctoral thesis, in that he could not account for the role that the magnetic field played in the optical spectra. More than a decade earlier, the Dutch physicist Pieter Zeemann had discovered that a particular frequency of light emitted would split in a magnetic field into two separate frequencies, whose differences were proportional to the magnetic field [31]. (Zeemann was fired for his efforts, as his postdoctoral advisor did not want him to work on magnetic field effects; fortuitously, he was awarded a professorship in Amsterdam for his work just in time to glory in the fame of his 1902 Nobel Prize.) There was no place in Bohr's theory for the magnetic field effect.

In truth, Bohr's "theory" consisted of a very few equations, with very great assumptions. He had avoided the decay of the electron energy through radiation by assuming that the quantization led to stability of the orbits. He made a crucial assumption on Planck's law that was inconsistent with the results he obtained. Thus, he had avoided difficulties by the assumption that they did not exist. Yet, his work was hailed as a great breakthrough. He didn't need to justify his results, reality existed in the measurements, and the measurements supported his breakthrough. But, in truth the theory had problems, as we have pointed out. This would be the case throughout his future—his work just wasn't quite right. Yet, through either scientific pressure or political pressure, this great Danish physicist received his Nobel Prize for this "seminal" work on the atom. Consequently, the Institute in Copenhagen became Valhalla for generations of young physicists who desired to work there, or at the least visit there, for a chance to sit at the feet of the great Niels Bohr. As Jones has pointed out [32],

> Germany at the turn of the twentieth century was definitely the place to be for aspiring scientists. Indeed, German was the language of physics. It's not that scientists weren't doing some good work at other universities in such places as England, France, and Austria, but ambitious young scientists headed for Germany to spend a year or two at the universities of Göttingen, Berlin, and Munich, enhancing their career opportunities by studying with the great men of German science.

From 1913, Bohr' fame would ensure that Copenhagen would become an additional site to visit, going beyond this list of German cities. Thus, generations of young physicists become indoctrinated to the philosophical views of Bohr.

Slowing for the Great War

World War I created a traumatic transition for the Europe of the nineteenth century. Although it started almost a decade and a half after the turn of the twentieth century, it was only the war that ended the lifestyle of the nineteenth century. As Stephen Brush has said [33], "following the usual convention in history, the First World War is taken as marking the end of the nineteenth century."

While scientists in some countries continued to work, a great many found themselves in the military, while others such as Moseley were lost to science, mere casualties of the conflagration. It had been a long time since the continent was engulfed in war, basically almost a century. Certainly, there had been conflicts. Prussia fought relatively brief wars with Denmark, Austria, and France around 1870, and Britain had their conflicts in South Africa. But by and large the century was peaceful, and there was nothing compared to the destruction brought by the Great War. However, Scandanavia and Holland remained neutral, although Denmark's neutrality was not quite true neutrality. Their mining of the waters around Denmark and the entrance to the Baltic was not viewed favorably by Britain and her allies. Nevertheless, within reason, life was not traumatized to the extent that the rest of the continent suffered. Hence, Bohr could carry on with his work and with the founding and building of the new institute. Moreover, students from abroad could travel to Copenhagen far easier than to the German universities. In addition, Bohr could easily travel to Britain, as he did several times during the war.

Most of the continent was dramatically shaken by the war. Countries were modified and new countries appeared as the map of Europe was redrawn by the victors. The Austrian-Hungarian empire disappeared and Germany was truncated. The Weimar republic was considered to be a disaster by some, especially with the hyperinflation of the early twenties. Yet, Germany itself had not suffered physically from the war. Certainly, it had suffered economically, but many wondered how they had lost the war, when no foreign army had come to Germany. It was a natural prescription for the fomenting of dissent, which the hyperinflation did nothing to dampen such views. One culmination, which would have tragic repercussions in later years, would be the beer hall putsch in later 1923. Suppression of the putsch only delayed the rising tide.

The universities tried to return to their life before the war. But there were differences. The economy was very different, and there were new pressures after the loss of population in the war. While superstars, such as Einstein and Planck, didn't have to spend time lecturing and supervising students, lesser faculty like Max Born (to be discussed below) were pressed to take on larger numbers of students. As Philipp Frank put it [34],

> in this way, there arose what was known in Germany as the *Betrieb* (mill), where to all outward appearances no distinction was made between worthwhile ideas and trivialities.... An agreeable feeling of activity surrounded both teacher and students. They became so engrossed in this activity and industry that the larger problem that the partial studies were supposed to elucidate was often forgotten. The production of dissertations and papers became an end in itself.

This would have a visual effect on the rise of quantum mechanics, as the economic, political, and educational problems would have a dramatic effect on the intellectual climate in Germany.

The nature of this climate is thought to have had an important effect, which we will see in the next section. As Holton has put it [35], "it has been argued that the widespread interest among German scientists in abandoning the principle of causality in the early 1920s (despite Einstein's opposition) is closely related to developments in the German intellectual environment." Now, there are certainly arguments about just what we mean by causality, but it is usually accepted that two events separated in time, in a single system, are related through a cause and effect phenomenon. It is perhaps best summarized by a quote from Frank [36]:

> Classical physics understood by the law of causality the calculability of future states from an initial state: if the state of the world or of an isolated system is known exactly at one instant of time, then it is known also for the future time.

While this is a good definition, it has its caveats. First, in order to know the state at a future time, it is presupposed that one has the full and *exact* equations of motion, not merely an approximation. Secondly, it is required to know the *exact* state of the system at the "one instant of time." We know from the more modern mathematical theory of chaos, that two nonlinear systems, which satisfy precisely the same equations of motion but which have almost trivial differences in the initial conditions, can have exponentially diverging future behavior. Even still, these latter two systems are causal. Some degree of uncertainty about the exact equations or about the exact initial conditions means that you will not have the ability to know the future state with certainty.

While the war slowed scientific advances, there were still results and advances that would affect our story. Arthur H. Compton was an American scientist, born in Ohio. He attended Princeton University, along with his brothers. They were the first group of three brothers to obtain doctorates from Princeton. Arthur graduated in 1916, then spent a year at the University of Minnesota, finally joining Westinghouse. At Westinghouse, he mostly worked on the sodium-vapor lamp, but also worked on aircraft instrumentation during the war. Nevertheless, he still spent time on physics, and in 1919 he published papers arguing that the size of the electron was much larger than what classical physics would tell us [37]. He argued that the experimental scattering coefficient for X-rays and γ-rays from different masses in the targets was much smaller than the theory suggested. He suggested that a "hypothesis that the electron has a diameter comparable with the wave-length of the hard X-rays will account qualitatively for these differences, in virtue of the phase difference between rays scattered by different parts of the electron." In fact, he suggested that the radius of the electron was about 2×10^{-10} cm, some three orders of magnitude larger than the classical assumption. Here was evidence that there was likely a quantum effect with the electron that led to its size being considerably larger than the classical value. In 1919, Compton won one of the first two National Research Council postdoctoral fellowships to study abroad, and chose to go to the Cavendish at Cambridge University. Here, he worked with George Paget Thomson, the son of J. J. Thomson, and who will play another part in the story later. With Thomson, Compton worked on the scattering and absorption of γ-rays. In 1920, Compton returned to the US and took a position at Washington University in St. Louis—at 28, he was a chaired professor and head of the department of physics [38]. As indicated above, he was interested in the scattering of X-rays already when at Westinghouse. He continued this work at his new post and discovered that the wavelength of X-rays was increased after scattering off an electron [39]. This meant that the incoming X-ray had transferred energy to the electron during the scattering process. The scattered electron became known as the recoil electron. This meant that the X-ray photon and the electron interacted just as classical particles would, and confirmed once again that Planck's hypothesis was correct and that Einstein's interpretation of the photoelectric effect was also correct. This would become known as the Compton effect, and he would receive the 1927 Nobel Prize for his work. The Compton effect will be a key experiment in the coming few years.

Satyendra Nath Bose was a mathematical physicist from India. He was educated in Calcutta and became a lecturer in the physics department at the University of Calcutta. Here, he translated Einstein's papers on relativity. In 1921, he moved to the University of Dhaka, where he worked on quantum theory. Convinced that one should not start from classical physics, he worked out a theory for black-body radiation and Planck's relation without any reference to classical physics. Rather, he based it on a new set of statistics for identical particles, like electrons. His arguments were based upon phase space arguments and assumed that population of the states would be limited by what became known as the Pauli exclusion principle, but it is unlikely that he would have seen Pauli's paper at this time. In essence, Bose created what we now call quantum statistics. As happens often, his paper was not accepted, and he sent the manuscript directly to Einstein, who clearly recognized the importance of Bose's work. Einstein translated the paper into German, and submitted it to *Zeitschrift für Physik* [40]. But Einstein then wrote several papers on absorption and emission in which he used Bose's results without proper referencing to the papers. As a result, the new statistics are now called Bose–Einstein statistics. With Einstein's backing, the papers became recognized for their importance, and Bose was recruited to Europe, spending time in Paris with young Louis de Broglie and Marie Curie, and in Berlin with Einstein. Upon returning to Dhaka in 1926, with Einstein's recommendation in hand, he was able to overcome the lack of a doctorate and still become professor and head of the department. But Dhaka was to become part of the new Pakistan partition in 1947 (and still later Bangladesh), so he returned to the University of Calcutta. In spite of the importance of his work, Bose was never awarded a Nobel Prize.

An important point in the evolution of quantum mechanics was the fact that now Einstein had become the principal spokesman for the quantization of light. This is significant because by 1925, he was an internationally known and revered scientist. Not only was he recognized for his great work on relativity, but now he was also recognized for his work on light quanta. While Planck was still active, he was approaching retirement and apparently was willing to let the younger Einstein fight the battles. And, as we will see, some of these were epic battles.

A Step Too Far

Niels Bohr had a dilemma. He had worked hard to establish that reality existed only when the measurement was done, thus pushing the philosophy supporting this early version of quantum theory toward Mach's worldview. But Mach had railed against the idea of the atom as an incarnation of discontinuity in science. And here, Bohr himself had brought in Planck's discontinuity as a key element of his atom. To really establish the philosophy, and allow him to become recognized as the great philosopher he envisioned, he had to get rid of the discontinuity and reestablish continuous dynamics. He had started on this project already with his second, 1913 paper, where he began with quantizing the angular momentum [26]. With this approach, he did not have to assume anything from Planck's relation. Quantizing the angular momentum was not as serious as it could not be measured, and hence was not connected to the real world, but quantizing the light field was measurable. Still, he did not need to quantize the light field, because the frequency was determined by the transition between his energy levels, and did not rely upon Planck, although this overlooked the basic relationship between energy and frequency. There was another problem, and that was the transition of the electron from one energy level to another with the emission or absorption of one of Planck's photons. Rutherford had already questioned him about these transitions and just how the electron knew where it was supposed to go. The door was opened for him by none other than Einstein.

Einstein had worried about the absorption and emission of radiation since his 1905 paper. He began to view it as a probabilistic event very soon [41, 42]. Finally, in 1917,[d] he published his important theory of absorption and emission, and defined what are known today as the Einstein coefficients [43]. Here, he noted that a molecule/atom can only exist in a series of states, each of which had a unique energy level, which followed Bohr's theories. He assumed that "a molecule may go from state Z_m to a state Z_n and emit radiation with an energy $E_m - E_n$" with a probability A_{mn} per unit

---

[d]An earlier version of the paper was published in the *Mitt. Phys. Gesell. Z.* the previous year, presumably from a talk he gave.

time. He assumed that the reverse process would be characterized by an absorption process with probability B_{nm}. But then he made the jump to say that these were not proper inverse processes. The downward transition with probability A_{mn} was made in the absence of any radiation field, whereas the upward process with probability B_{nm} was made in the presence of an electromagnetic field, such as light. Hence, there must be an equivalent downward transition in which the molecule emits radiation *due to the presence of an electromagnetic field*. He said this emission process would be denoted by the probability per unit time B_{mn}. Hence, the B processes are stimulated by the presence of the electromagnetic field, while the A process could occur spontaneously. From this view, he could once again derive the Planck black-body radiation law. Importantly, while the spontaneous emission process could give off light in any direction, the stimulated emission process would give light moving in the same direction as the radiation field which induced the transition. This would become the basis of the laser a few decades later,[e] but the entire description provided by Einstein was that these processes were probabilistic in nature and the Planck law was a probabilistic law. The absorption and emission of radiation was a statistical process described by a well-defined probability distribution function.

Here, then, was an opening for Bohr. If black-body radiation itself was a statistical process, then perhaps Planck's formula $E = hf$ was also only an average quantity and not a true quantization. This is quite a jump of faith, but it would get rid of the annoying quantization and reestablish some sort of continuity. At the time, Bohr was working with two young scientists in Copenhagen.

(Hendrik Anthony) Hans Kramers was a young Dutch scientist who had come to Copenhagen to work with Bohr. He was a graduate student at Leiden under Paul Ehrenfest, who sent him to Bohr for a period to learn the new quantum theory from the "master." Kramers had wanted to go to Göttingen to work with Born, but this was denied him due to the war, so he chose Leiden instead.

---

[e] Properly speaking, the maser came first, where the maser (microwave amplification of stimulated emission of radiation) and laser (light amplification of stimulated emission of radiation) are actually acronyms which describe the process.

Thus, he did most of his doctoral work in Copenhagen under the direction of Bohr, but obtained the degree from Leiden. Returning to Copenhagen, he worked with Bohr and gained a position at the university. He then stayed with Bohr as his chief assistant in Copenhagen for a decade. Originally, he was a strong proponent of the quantization of light, but his view changed dramatically under pressure from Bohr. If we are to believe Pais and others, we have the following [44]:

> The indications are that Kramers had obtained the conservation law description of the Compton effect [inelastic scattering of photons by electrons], using the photon description explicitly . . . Bohr would object violently to the publication of these results. . . . Kramers . . . was simply ground down by Bohr.
>
> After these discussions, which left Kramers exhausted, depressed, and let down, . . . sick . . . and spent some time in the hospital . . . he soon became violently opposed to the photon notion.

John Slater was a young American who, after finishing his doctorate at Harvard in 1923, decided to make the pilgrimage to Europe, with the aid of a traveling fellowship from Harvard. He briefly went to Cambridge, but finally came to Copenhagen as a young postdoctoral scholar to work with Bohr, but prior to coming to Copenhagen, he had published a paper with the goal of providing a more adequate picture of optical phenomena [45]. In this, he assumed that Compton was correct in his scattering picture, but supposed that [46]

> each atom should be supposed to "communicate" constantly with all the other atoms as long as it is in one of its stationary states, this communication being established by a virtual field of radiation. . . . The part of the field originating from the given atom itself is supposed to induce a probability that the atom loses energy spontaneously, while radiation from external sources is regarded as inducing additional probabilities that it gains or loses energy much as Einstein had suggested.

With these two young scientists, Bohr formulated the paper that came to be known as BKS, after the initials of the authors [47]. In this paper, Bohr proposed doing away with conservation of energy, conservation of momentum, and causality! Here, the authors say that "it does not seem possible to avoid the formal character of the quantum theory which is shown by the fact that the interpretation of atomic phenomena does not involve a description of this mechanism of the discontinuous processes." It was clear that they were going to address the issue of the time scale of the transition between energy levels. The source of the transition was not to be a spontaneous process, but an induced process, as they remarked: "The essentially new assumption introduced . . . that the atom, even before a process of transition between two stationary states takes place, is capable of communication with distant atoms through a virtual radiation field." Then, they provided a direct disrespect of Planck with the comment "the theory of light quanta can obviously not be considered as a satisfactory solution of the problem of light propagation.

The authors return to the issue of the time scale of the transition and open a new door, with the comments [43] "great difficulties have been involved in the problem of the time-interval in which emission of radiation connected with the transition takes place . . . whether the detailed interpretation of the interaction between matter and radiation can be given at all in the terms of a causal description in space and time of the kind hitherto used for the interpretation of natural phenomena." As Jammer puts it [48], "it was the first major paper in physics which deliberately and programmatically renounced methods of explanation as well as fundamental principles of classical theory." They return to this later with [43]. "We abandon on the other hand any attempt at a causal connexion [*sic*] between the transitions in distant atoms, and especially a direct application of the principles of conservation of energy and momentum, so characteristic of the classical theories." And, if the reader missed this point, they reiterate it and "assume an independence of the individual transition processes, which stands in striking contrast to the classical claim of conservation of energy and momentum. . . . This independence reduces not only conservation of energy to a statistical law, but also conservation of momentum." Finally, they return to the duration of the transition: "the duration of which we shall assume at any rate not to be large compared with the period of the corresponding harmonic motion . . . and [governed by] the time-interval in which the atom will be able to communicate with other atoms." They applied this theory to Compton scattering and proposed that this should be a random process, in which "the illuminated electron possesses a certain probability of taking up in unit time a finite amount of momentum in any given direction." That is, momentum would not be conserved in Compton scattering.

Now, was this the view of all three of the authors? Not quite! We have already seen that, while Kramers had begun with belief in the quantization of light given by Planck, he had been ground down by Bohr's consistent pressure to conform to Bohr's new ideas. Slater also later said [49]:

> The idea of statistical conservation of energy and momentum was put into the paper by Bohr and Kramers, quite against my better judgement.... I wished to introduce probability only so far as the waves determined the probability of the photon's being at a given place at a given time. Bohr and Kramers opposed this view so vigorously that I saw that the only way to keep peace and get the main part of the suggestion published was to go along with them.

More to the point, Slater said [50]:

> In other words, he [Bohr] wanted to make the whole thing just as vague as he could. Kramers was always Bohr's "yes-man" and wanted to do exactly the same thing . . . those papers were dictated by Bohr and Kramers very much against my wishes. I fought with them so seriously that I've never had any respect for those people since. I had a horrible time in Copenhagen.

Even before going to Copenhagen, Slater had concerns about Bohr's [50]

> hand-waving approach to anything. I had supposed, when I went to Copenhagen, that although Bohr's papers looked like hand-waving, they were just covering up all the mathematics and careful thought that had gone on underneath. The thing I convinced myself of after a month, was that there was nothing underneath. It was all just hand waving.

The BKS paper was written very shortly after Slater's arrival in Copenhagen. While he would remain for several more months, he spent most of this time at an island spot belonging to his landlady, where he could work effectively alone.

Here, we have the essence of Bohr. His papers contain little proper theory, and much "proof by intimidation," and now we have almost physical, and certainly mental, harassment of his colleagues, in an effort to enforce his point of view. This will not be the sole instance of this, as many others encountered this behavior upon visits to Copenhagen. As Boya succinctly put it [51], "The degree of brainwashing by Bohr in all of us is remarkable (I learned the expression from Murray Gell-Mann)." In fact, what Murray Gell-Mann had said, writing some 50 years later, is [52]:

> The fact that an adequate philosophical interpretation [of quantum mechanics] has been so long delayed is no doubt caused by the fact that Niels Bohr brain-washed a whole generation of theorists into thinking that the job was done 50 years ago.

The BKS paper was met with a resounding thud. Many thought it difficult to read due to Bohr's ability to obfuscate important points, yet he also tended to leave no chance to misread his important points. However, it made its way into the popular press, and journalists had, of course, to interview the main important scientists. By this time, the principal spokesman for light quantum was Einstein. In the end [53], "the dispute between Bohr and Einstein was now in the public domain, with sides clearly drawn." And Einstein would win this round. Experiments done by Walther Bothe and Hans Geiger showed the assumptions of BKS were wrong—energy and momentum remained conserved [54, 55]. Bothe had developed the coincidence method, in which the scattered X-rays would go through one detector, while the electron recoil would go to a second detector. The experiment was arranged so that the output pulses of the two detectors were approximately coincident in time. This raised the sensitivity of the measuring apparatus where the noisy output of one detector could be correlated with the noisy output of the second detector, reducing the overall signal-to-noise ratio. He and Geiger applied this to the Compton effect with very precise results that made it absolutely clear that energy and momentum were conserved. Bothe would receive a Nobel Prize in 1954 for the development of the coincidence method. Thus, within a very few months, BKS had clearly been demonstrated to be wrong, but this apparently did not deter Bohr, and he still had causality at risk. We defined classical causality above, but we can say more about it. Cushing gives a nice perspective [56]:

> While locality is a constraint we have traditionally imposed on our theories, we have at times (as with classical gravitational theory) been willing to yield on it. But even then we have required a theory that allowed a causal story to be maintained in terms of interactions among parts of a system. This has enabled us to picture, in some sense, what is going on.

In essence, this allows a deterministic time evolution of the system. Bohr would have none of this. If reality were to be determined by the measurement, then there could be no underlying process which described this reality. As McEvoy puts it [57], "Bohr . . . moved even further in the direction of rejecting the possibility of a space-time description after the failure of the BKS theory."

Kramers would return to Holland and become a professor in Leiden, while Slater returned first to Harvard and then to the Massachusetts Institute of Technology, where he established the foremost group on the quantum theory of electronic structure. Most modern studies with density functional theory probably have their roots in the work of Slater. Upon mandatory retirement from MIT, he moved to the University of Florida to continue his work.

In the end, Bohr had not solved his dilemma. He had achieved considerable fame and recognition for his atom papers, but he wanted more. He now had a public dispute with Einstein, who was being talked about as the greatest physicist of the age, but young scientists were flocking to Copenhagen, not to Berlin. They came to him, not to Einstein. Why shouldn't he be regarded as the best? To do so, he needed more than his atom papers. He thought BKS would be another great breakthrough and people would hail him for his great insight, but it was a dud. Fate would be kind to him, though, and he would get another chance to be the defender of the faith. In this effort, he would continue to attack causality. If one really believed that reality occurred only when the effect was measured, then he felt that a statistical theory of quantum effects needed to be free from causal dependences. He had tried and failed to get rid of energy and momentum conservation. Getting rid of causality could provide another feather for his hat.

The Rise of Duality

There remained problems in Bohr's atomic model. A key element that came from his formulation was the quantization of the angular momentum of the electron as it orbited the nucleus. There was no reason for this to occur; it had originally come from his assumption of Planck's law and fell out of the "theory." Even later, when he began with this assumption, he had no deep mathematical rationale for this assumption. If there were some underlying reason for this effect, it had not come to Bohr, but then his success lay not in the mathematics but in the fit it made to the atomic spectra that was experimentally observed. With this view, explaining the quantization was not important to him. It is not surprising that the answer came from elsewhere.

Louis Victor Pierre Raymond de Broglie was born into French nobility, the son of the fifth Duke de Broglie, and he led a life of relative ease. His early interests were in the humanities and his undergraduate degree was in history. Afterward, probably under the influence of his older brother Maurice (who would become the sixth Duke de Broglie), Louis turned his attention to mathematics and physics and received a second undergraduate degree. When the first world war began, Louis was 21 and of prime age for the army. Here, the family connections were crucial and Louis served his time in Paris as a telegraph operator in the station at the Eifel tower [58]. Maurice had a strong interest in physics, and maintained an extensive laboratory in the basement of the family home. He thus was known to important physicists in Paris, and this would allow him to attend the first Solvay conference as secretary to Marie Curie and Paul Langevin (discussed in a later chapter). And, in turn, he brought along little brother Louis to observe. The younger

Louis would pore over the notes that Maurice made on each day's proceedings. This, and the existing laboratory at home, probably helped Louis in his decision to pursue physics and mathematics. The exciting new physics interested him, and Louis published many papers on light propagation and quantization. In fact, he jumped on the results of Taylor (previous chapter) on the particle and wave aspect of light as it created the interference pattern (where each photon creates one spot on the screen, or photographic plate, and the interference pattern came from an ensemble of photons). This led him to the early version of the guiding wave principle, in which the corpuscles (he referred to them as atoms of light) rode on a nonmaterial light wave of frequency f [59]. He would come to refer to the wave guiding the particle as the "pilot" wave. He had been attracted to Bohr's "theories," but usually took a different philosophical view as was clear from this paper and his later work. Nevertheless, he had been noticed by the Copenhagen crowd, which wasn't particularly pleased with his ideas. Then, as he worked on his doctoral thesis, it occurred to him that if light were both a wave and a particle, then a particle such as the electron should have the same properties, that is waves and particles should be duals. As Cushing puts it [60]: "De Boglie's solution to the wave-particle duality was a synthesis of wave *and* particle, versus the wave *or* particle of the eventual Copenhagen interpretation." As de Broglie said later [61],

> When I started to ponder these difficulties two things struck me in the main. Firstly the light-quantum theory cannot be regarded as satisfactory since it defines the energy of a light corpuscle by the relation ... which contains a frequency ... a purely corpuscular theory does not contain any element permitting the definition of a frequency. This reason alone renders it necessary in the case of light to introduce simultaneously the corpuscle concept and the concept of periodicity.
>
> On the other hand the determination of the stable motions of the electrons in the atom involves whole numbers, and so far the only phenomena in which whole numbers were involved in physics were those of interference and of eigenvibrations. This suggested to me that electrons themselves could not be represented as simple corpuscles either, but that a periodicity had also to be assigned to them too.

As a result, the quantization of the angular momentum was clear, for if the electron was a wave, then the orbit around the nucleus must involve a distance that was a multiple of the wavelength of the wave (his periodicity). Hence, it had to be quantized, since only a very particular momentum and energy would allow the wave to fit into the circumference of the orbit with this property. This led to his famous formula

$$\lambda = \frac{h}{p}, \tag{2.8}$$

where p was not the momentum of the photon, but the momentum of the electron in its motion around the nucleus. He finished his thesis and submitted it to the university, in essence to Paul Langevin.

In truth, de Broglie's formula above is precisely the same as for a photon (which he of course knew), but the momenta are different. In the case of a photon, Einstein had already shown that the photon had a momentum hf/c [43]. So, Planck's law could be rearranged using $f = c/\lambda$ as $p = hf/c = h/\lambda$, which is the same as Eq. 2.8. The difference in the two cases is the momenta, as in the photon case the momentum is a linear function of the energy. For a classical particle however, the energy and momentum were related by $E = m_0 v^2/2 = p^2/2m_0$, where $p = m_0 v$. Hence, in the case of the particle, the momentum varies as the square root of the energy and also depends upon the mass. For a 1 eV photon, we find that the wavelength is 1.24×10^{-6} m. For an electron with an energy of 1 eV, we find the wavelength is only 1.16×10^{-9} m, which is more than three orders of magnitude smaller. The fact that the particle wavelength is so much smaller explains why the particle waves had taken so long to be recognized.

There are many views about what Langevin thought about the thesis, ranging from distrust of the young de Broglie to not understanding the thesis himself. Whatever the reason, Langevin sent a copy of the thesis to Einstein. Einstein was taken with the ideas and said so, following which de Broglie was awarded his doctorate. As Jones as said [62],

> What intrigued Einstein was that de Broglie had made a direct link between matter, waves, energy and the quantized atom. He had taken the properties of a particle (energy and momentum),

married them to the properties of a wave (frequency), and plugged in Planck's constant h, *et voilà*, de Broglie had produced a simple formula to show that all matter could be considered as a wave. This was something Einstein could work with, and he did.

A part of his thesis, which included the ideas of matter waves, was published in the UK [63]. With this work, waves and particles were to be considered as two interpretations, or observations, of a single quantity. It would become known, mostly through the views of the Copenhagen crowd (of which he was not a member), as wave–particle duality because the decision as to which appeared would arise in the experiment. In this way, de Broglie's seminal ideas would become warped for someone else's greater good.

As de Broglie's work became known, he was hailed for his advances and he would join the list of Nobel Prizes for advances in quantum theory: Planck in 1918, Einstein in 1921, Bohr in 1922, and de Broglie in 1929. Yet, his work was not acceptable to all, and there would be strong debates in quantum community that have lasted until the present time.

But, were the electrons really waves. If so, they should show diffraction and interference. Already in 1921, Clinton Davisson and Charles Kunsman, working at the AT&T research laboratories in America, had noticed that the scattering of electrons from Ni and other metals showed unexpected peaks in the angular variation of the scattering [64, 65]. In sending a beam of electrons into a crystal, such as Ni, many effects can be seen. If the energy of the entering electrons is sufficiently large, they can knock other electrons out of their atomic orbits and these electrons can be emitted in complete analogy to the photoelectric effect, with the incident electron replacing the photon. But Davisson and Kunsman also observed electrons with almost the same energy as the incident beam being reflected from the crystal over a large range of angles. Studies of the angular distribution of the emitted/scattered electrons showed the appearance of peaks much like diffraction peaks, apparently due to the periodic nature of the atomic arrangement. Others observed similar effects in the scattering of low energy electrons from gasses [66]. In these studies, the scattering cross-section had an oscillatory behavior with peaks and valleys as the energy was varied as in the case with Ni. Walter Elsasser proposed that these effects were the result of the diffraction of electron waves [67], and his paper was published only after the editor consulted with Einstein [68]. Just after this, George Thomson, the son of J. J. Thomson, obtained clear diffraction patterns of electrons striking thin metal films [69]. Sending the electrons (which they called cathode rays still) through the thin films, they observed the diffraction rings of the exciting electrons. Thus, it was finally clear that de Broglie was absolutely correct. The electrons in Bohr's atom could act as waves and exhibit wavelike features. Davisson and Thomson would finally receive the Nobel Prize for their work in 1937. George Thomson's father had received the Nobel Prize for showing that the electron was a particle; the son would get the prize for showing that the electron was a wave. This must have stimulated interesting family discussions.

Thus, we had the complete duality established. Light waves were composed of particles as Planck had established in 1900. This had been confirmed by Einstein with his explanation of the photoelectric effect. Electrons, and likely other particles, were waves and exhibited wave diffraction and interference. So waves were particles and particles were waves. Indeed, the world was a strange place now. Quantum phenomena had given new meaning to physics. The final act of duality would come when electrons were shown to give the same results in the two-slit experiment as photons (a video of the electron version has been created by Akira Tonomura of Hitachi [70]). But, de Broglie had given us another problem. If electrons were waves, where was the equation which governed their motion? Light waves had Maxwell's equations, which provided a litany of governing equations. So far, electron waves had zero equations.

The New Breed

By 1920, it became a rite of passage for young physicists to pass through Copenhagen. There were many who passed through in the first half of the 1920s, after Bohr had gotten his Nobel Prize. But there were others who had become prominent, but did not pass through Copenhagen, of which Louis de Broglie was one. And there were still others who went briefly to Copenhagen and were repelled at the viewpoints of Bohr. It would become clear that the philosophical differences between those who went to Copenhagen, and those who did not, would be dramatic. A few of all three of these groups would play prominent roles in the new quantum mechanics that would arise during the second half of this decade. It is this so-called younger generation that we want to introduce here, in no particular order, as we will encounter them in more depth in the next chapter.

It is hard to think of Max Born as a youngster—he was actually three years older than Bohr and three years younger than Einstein, but his play in the quantum world was still ahead of him, and he would prove to be an attractive force for young talent equal to that of Bohr. Born was born in Breslau, which is now part of Poland and known as Wroclaw, in 1882. He was educated at the University of Göttingen,[f] the center of mathematics in both Germany and most of the rest of Europe. Here were the great mathematicians David Hilbert, Felix Klein, Richard Courant, and Hermann Minkowski. Hilbert may have been the greatest and is known for his 23 problems. In 1900, at an international conference in Paris, Hilbert talked on unsolved problems in mathematics and provided a list of 23 that he considered the most significant. This became a roadmap for future mathematicians, and some of them have still eluded closure. Hilbert also recognized potential in Born and became his mentor. Intellectually, Minkowski was probably Hilbert's equal and is known for providing the mathematical framework upon which relativity is based. He wrote a paper on elasticity, a topic suggested to him by Klein, but elected to do his thesis on a more mathematical stability solution for the elasticity paper under Carl Runge. This work led him to receive a university prize for it. Following Born's graduation in 1904, he began working with Minkowski and the theory of relativity. He still had to cope with the compulsory military service of the time, which he had deferred as a student. Nevertheless, upon graduating, he found himself in the army with a regiment stationed in Berlin. Bad and good often come together, and for Born,

---

[f]In Germany and Austria, the first university degree was the doctorate, as there was no tradition of a "bachelor's" degree. Prior to becoming faculty, however, they had to achieve the habilitation, a higher degree based upon their research.

this was his asthma. Following a severe and disabling attack, he was released from the army in 1907. This allowed him to take a fellowship at Cambridge, where he could work with Thomson and his colleagues for six months. Upon return to Germany, the army decided that perhaps his asthma wasn't all that bad, and he again found himself in uniform until another severe attack set him free again. By this time, he had learned about Einstein's relativity theory, and sent some of his own ideas to Minkowski, who encouraged him to return to Göttingen, which he did. Here, the mathematician Otto Toeplitz helped him with matrix algebra so that he could more easily work with the Minkowski four-dimensional space. Minkowski was interested in coupling relativity with electrodynamics, and Born began to work with him on this area. Unfortunately, the sudden death of the former ended the collaboration. However, Born was able to complete the development of the theory, and published the results [71]. Shortly after this, he finished his habilitation on Thomson's theory of the atom [72]. He continued to work in Göttingen, and by the end of 1913 he had published more than two dozen papers and had gotten married. In early 1914, he was called to Berlin by Max Planck to take a new professorship in theoretical physics. The new chair had been offered first to von Laue, who had turned it down, but soon had second thoughts. Nevertheless, the war had broken out, and Born returned to the army. Fortuitously this time, he was assigned as a radio operator in the signal corps, but he was soon drawn into research on the use of sound for target ranging, a subject we know better today as sonar. By now, Einstein was also in Berlin, and the two became lifelong friends.

For various reasons, Born went to Frankfurt rather than Berlin after the war. Then, in 1921, he returned to Göttingen as professor of physics. Here, his goal was to raise the stature of the mathematical physics program to the level of the mathematics program. This would prove fruitful to the advances of the coming quantum revolution, with many of the new stars passing through Göttingen. As part of the plan to raise the status of physics in his new university, he arranged to bring Niels Bohr to Göttingen to give a series of lectures, which was termed a Bohr fest, and he invited many people to attend. Sommerfeld came and brought along some of his students, one of whom was Werner Heisenberg. Paul Ehrenfest came. Wolfgang Pauli was there already, spending a year in Göttingen after finishing his own dissertation. Heisenberg was elated to meet the famous Bohr, and spent considerable time in discussions with him. Bohr even invited Heisenberg to Copenhagen, but things didn't work out well at this time for such a visit. Sommerfeld was preparing to take an extended trip to the United States and took the opportunity to arrange for some of his students to stay in Göttingen during his absence, another fortunate break for the young Heisenberg. Born's efforts to raise the status of physics were paying off, and Göttingen would be another stop for young scientists seeking wisdom by travel. In addition to those already mentioned, others among the young who would be attracted to Göttingen and Born's new Institute of Theoretical Physics were Pascual Jordan and Paul Dirac. As the new quantum mechanics unfolded, John von Neumann would arrive as an assistant to Hilbert. The group around Born would be the center for development of the principles of the new quantum mechanics. As much as Born was interested in theoretical physics, he still was a firm believer in the importance of experimental work to test, and help in the development of, new theories. In arranging to come to Göttingen, he had also arranged a new position for experimental physics to be filled by his friend James Franck, who was already known for his experimental work in Berlin.

Wolfgang Ernst Pauli was an Austrian, born in Vienna, to a well-to-do family (his father was a chemist). His middle name is a tribute to his godfather, Ernst Mach! Pauli's father had gone through school with Mach's son, and the families were close. Pauli's brilliance was recognized at an early age, especially when he published his first paper, on a form of relativity theory discussed by Herman Weyl, only months after finishing gymnasium (nearly equivalent to a high school in the USA) [73]. Pauli then went to Munich so that he could do his university work with Sommerfeld. He finished his doctoral work, on the quantum mechanics of diatomic hydrogen, in 1921, and then went to Göttingen for a year to work with Born, and then went to Copenhagen to spend a year with Bohr. Thus, Pauli was in Göttingen at the time of the Bohr fest, and this probably helped influence his decision to spend the next year in Copenhagen. It was at the Bohr fest that Pauli met Paul Ehrenfest, apparently for the first time. Pauli was rather rude, and Ehrenfest is supposed to have said, "I like your publications better than you," to which Pauli replied, "Strange, my feeling about you is just the opposite" [74]. Following his stay in Copenhagen, he tried returning to Göttingen, but found that he preferred thinking about the physics rather than focusing on the mathematics that Born preferred. Hence, he took an opportunity that arose and became a lecturer at the university in Hamburg in 1923. In 1928, he became professor of physics in Zürich, and eventually became a Swiss citizen. Pauli was a man in the mold of Bohr, as it is sometimes said that any theory which he had not already thought of was simply rubbish.

In 1921, Pauli finished a long article on relativity theory [75], as apparently his interest in this topic had continued since his high school days. He was already becoming a Mach proponent, even before his detour to Copenhagen. In this article, he says [60], "There is no point in discussing . . . quantities [that] cannot, in principle, be observed experimentally." Hence, if one could not see the quantity via experiment, he felt that they were [75] "unobservable by definition, and thus be fictitious and without physical meaning." It was thus already clear that he would naturally be pulled into Bohr's camp and point of view. Although he would try to maintain his independence, his views would lead to his being closely associated with Bohr. Pauli did not particularly like the wave *and* particle approach that de Broglie introduced with his pilot waves in 1924, and would forcefully express these objections a few years later at the Solvay conference (which we discuss in Chapter 5). While in Hamburg, he addressed the problem of the Zeeman effect and proposed that each quantum state possessed a new state which had two values [77], conveniently corresponding to the splitting of each atomic spectral line that Zeemann had discovered. During a seminar, the American-German Ralph Kronig, a student at Columbia, but in Europe on a travel fellowship at the time, asked him whether this could correspond to the electron spinning on its own axis. Pauli dismissed the idea in a manner, which may be suitably expressed as unworthy of consideration. The next year, George Uhlenbeck and Samuel Goudsmit, working with Paul Ehrenfest in Leiden, identified the two states as the spin angular momentum of the electron. Unfortunately for these latter scientists, Kronig's unpublished suggestion was already known in the scientific world and no one would receive recognition for this important discovery. We will return to this in the next chapter.

Werner Heisenberg was born in Würzburg, a wonderful town in northern Bavaria famous for a particular type of white wine. His father was a gymnasium teacher at the time, but would become the only professor in Germany (in Munich) in the area of medieval and modern Greek studies. It appears that Heisenberg was raised in an upwardly mobile family that valued education. With his family now in Munich, it was natural that he would attend the university in Munich, where he studied physics with Arnold Sommerfeld. However, students would often attend several universities to broaden their exposure to various fields. For Heisenberg, this meant spending time in Göttingen, where he could study mathematics with Max Born and David Hilbert. As mentioned above, Sommerfeld took Heisenberg to Göttingen for the Bohr fest in 1922, and arranged for several of his students, including Heisenberg, to spend time with Max Born while he went off for a trip to the United States. Heisenberg had met Pauli earlier, as the latter was only a couple of years ahead of Heisenberg in Sommerfeld's group. Pauli would be his tutor and lab instructor, and became something of an "older brother" to the younger Heisenberg. They renewed their acquaintance and friendship in Göttingen at the Bohr fest, and Pauli would remain a kind of advisor for Heisenberg for years afterward. Heisenberg returned to Munich to finish his doctorate in 1923, on the topic of turbulence in hydrodynamics. But this was a near thing. In Munich at the time, the oral examination involved two professors and it was generally felt that being a theoretical physicist was not an adequate education. The examination committee included Willy Wien, who felt that even the brilliant theoretical students of Sommerfeld needed to know something about experimental physics [78]. It turns out that Heisenberg was an exceptionally poor experimentalist, and thus could not handle the questions about experiments on his final examination. Heisenberg stumbled during the final oral, and an argument arose between Wien and Sommerfeld. The former wanted to fail the young man, but in the end a compromise grade of III was given, which is equivalent to a C, barely passing. To say that Heisenberg was mortified would be an understatement. He had already arranged to return to Göttingen to work with Born, but now he had to inquire whether Born still wanted him with this disaster. Born had his own problems with experimental work, so that he fully understood the problem and the politics. He encouraged young Heisenberg to return to his fold. As we will see later, however, as a theoretical physicist, Heisenberg was not as prepared in mathematics as he should have been. The next year, he received a Rockefeller grant to study in Copenhagen with Bohr, where, he apparently received his indoctrination to Bohr's view of the world.

In his first stint in Copenhagen, Bohr clashed badly with Kramers. Kramers was the lead assistant in the Bohr group and valued his status as the leader of the young members. It was clear to any looking at the group that Bohr was surrounding himself with bright young scientists. Even among this group, Heisenberg appeared to stand out, and so was a direct challenge to Kramers maintaining his position. As could be expected, Kramers was working on a theory of optical scattering from the quantum atom. Heisenberg was interested in the work, but thought it lacked a proper mathematical base (and suggested he could provide it and be a co-author). Heisenberg had been working with Born on this topic before coming to Copenhagen, and properly felt he had some ideas to contribute to the work. As Sheila Jones puts it [74],

> It was a clash of personalities, but also a clash between the visualizability of the orbital model of the atom, championed by Kramers, and the non-visualizability of mathematical symbolism, put forward by Heisenberg ... the two finally took their dispute to Bohr ... [who] ruled that Heisenberg's mathematical approach would be used and that his name would go on the paper as co-author.

Their paper was finally published in early 1925 [80], just as Heisenberg was preparing for his return to Göttingen, and his next burst upon the world stage. Kramers would leave Copenhagen and return to Leiden the next year.

Interestingly enough, in his paper with Kramers, Heisenberg writes about "a wave-theoretical analysis of the scattering effect of an atom." Within little more than a year, he would be rejecting wave theoretical approaches to quantum mechanics vehemently. This paper also was based upon the premises put forward in the BKS paper, and the authors would propose that light incident upon an atom would lead to the emission of different frequencies, which differed from the incident frequency by the frequencies of the allowed atomic transitions. Hence, they were proposing a nonlinear interaction within the atom, but the impact was to be cut short by the new quantum mechanics which would arrive within a few months.

Pascual Jordan came from a distinguished family in Spain (the name had been Germanized from Jorda when his father had moved to Hanover. He began his education in Hanover, but moved to Göttingen to complete his doctorate with Courant and Born. At the time, Jordan was a busy beaver, in that he was involved in transcribing lecture notes into a proper manuscript for a book being written by Courant and Hilbert. In addition, he was also working with James Franck on another book and working on his dissertation. For the latter, he was trying to bridge the gap between Einstein's light quantum and Bohr's new BKS paper (which was trying to do away with the light quantum). As a result of his work load, he was fluent not only in physics, but also in mathematics and this would be crucial to the developments that would flow from Göttingen. But, he had a problem, a speech impediment, which kept him from the major conferences and would probably affect his gaining the recognition he deserved in the evolution of quantum theory. Jordan had another problem in that he was an ardent national socialist well before the rise of Hitler. This was, after all, the 1920s at about the time of the Munich beer-hall putsch. While Born had converted, he still had a Jewish past, as did Bohr, Einstein, and many other prominent scientists. Nevertheless, Jordan worked well with Born and the other youngsters coming to Göttingen, but this would fall apart soon enough.

And then there were the outsiders, the first of whom was Paul Adrien Maurice Dirac. His father was a Swiss immigrant to the United Kingdom, and Paul received his undergraduate education in Bristol in electrical engineering. While at home, his father enforced speaking in French, which probably led Paul to develop a demeanor that included limited speaking and some withdrawal from the action around him. He was admitted to the University of Bristol at 16 and graduated at 19, remarking later on his engineering education [81], "I think that this engineering education has influenced me very much in making me learn to tolerate approximations." While he was admitted to Cambridge, the cost was beyond the limited means of the family, so Paul pursued a degree in mathematics in Bristol. Again, he took the entrance exam to Cambridge and did well enough to receive a fellowship that was sufficiently large to cover his costs. Cambridge was a new environment for him, especially the modern fields of research, and he began his work with Fowler. He remarked [80], "When I went to Cambridge, I learned about the Bohr theory of the atom from Fowler. I had no idea that atomic theory was so developed. It came as a surprise to me. I had not heard of the Bohr theory of the atom at all in Bristol." With Fowler, he pursued research on relativity theory at Cambridge, switching to quantum mechanics only at the very end of his studies. Nevertheless, he managed to publish a few papers on relativity and statistical physics. Interestingly enough, one of these papers connected the Doppler effect with Bohr's frequency condition [82]. He received his doctorate in 1926, with the first thesis on quantum mechanics that had been awarded anywhere. The following year he became faculty at Cambridge as a Fellow of St. John's College, but he would make the required pilgrimage to the continent to Copenhagen and then to Göttingen. But, as we will see, by this time he was already playing on the world stage. He would be elected a Fellow of the Royal Society in 1930, at the age of 28. In 1932, he was appointed to be the Lucasian Professor of Mathematics at Cambridge, a position which Newton had taken just over 260 years earlier. He would share the 1933 Nobel Prize with Erwin Schrödinger. Following his mandatory retirement at Cambridge, he moved to Florida State University. Florida may have Slater, but Florida State had Dirac.

Like Max Born, Erwin Schrödinger was older than the other members of the new breed. Erwin Rudolf Josef Alexander Schrödinger was only two years younger than Niels Bohr. He father was a botanist and his mother the daughter of a professor of chemistry. But, his mother was half English; his maternal grandmother coming from an upper level family in Britain. Young Erwin would spend many summers in Britain with this family, and this would allow him to learn English prior to any schooling [83]. This led to his being raised as a Lutheran, there being no Church of England in Austria, although he gave up formal religion at a relatively early age. He completed his dissertation in 1911 and became an assistant to Franz Exner and Friedrich Hasenöhrl. During this period, he became very interested in color theory, which detailed how vary parts of the visible spectrum combined to produce different colors. He completed his habilitation in 1914, just in time for the war. Schrödinger spent the war as an officer in the fortress artillery, and a good bit of the time in the Italian alps with his unit. Following the war, he took up a position with Max Wien (who should not be confused with Willy Wien), inventor of the Wien bridge, in Jena, but quite soon received a faculty position in Stuttgart. A year later, he moved to Warsaw as a full professor, and then quickly to Zürich. It was clear that he was willing to move rapidly to advance his career. He would stay in Zürich for several years, until being called to Berlin, to succeed Max Planck, in 1927, which was after his papers on the wave mechanics formulation of the new quantum mechanics.

Schrödinger was married in 1920. Perhaps the most unusual aspect of Schrödinger and his wife was what we could call today as an "open" marriage [84]. It's said that Erwin seldom observed a woman without developing a significant attraction, and is known to have had more than his fair share of liaisons, but the marriage was a true two-way street, and his wife also had some affairs. When he would leave Austria in 1938, due to the Anschluss, he went first to Oxford, but his unusual living affair (with two women at the same time) did not sit well, and moved to Dublin, where the Irish were willing to overlook his peculiarities. He settled in well there with both his wife and his mistress. Surprisingly, he and his wife stayed together until his death.

Schrödinger's early work concerned dielectrics and then work on Einstein's gravitation. In the earlier 1920s, he produced a series of papers on color theory, for which he became rather well known. Perhaps not surprisingly, he also discussed the Doppler effect in connection with Bohr's model of the atom [85]. In fact, it was this paper that stimulated Dirac's later paper [81]. As Dirac put the problem [81],

> When an atomic system emits or absorbs radiation, the frequency of the radiation is connected with the change of energy of the system by Bohr's frequency condition $\delta E = hf$ ... the frequency of radiation being measured in a frame of reference in which the atom as a whole is at rest.

Now, as Dirac points out, apparently Bohr was aware of this but didn't seem to understand how to take the motion of the atoms into account. First Schrödinger, and then Dirac, showed that it was a normal result of relativity theory. Again, as Dirac stated [81],

> The question has been considered by Schrödinger on the basis of Einstein's hypothesis that an emitted quantum of energy ... carries an amount of momentum hf/c.
> The law of transformation of f is the same as that given by the ordinary Doppler principle.... Provided the emitted radiation is entirely directed, consistent results will be obtained whichever frame of reference is used for applying the frequency condition.

Hence, the moving atom merely provides a frequency shift consistent with its motion. In the case of a gas, however, the motion of individual atoms has random directions, and the frequency shift has a dependence on the angle between the direction of the motion and the direction of the observer. Overall, this will provide an apparent broadening of the spectrum due to this random motion. Such broadening makes the observation of splitting of individual spectral lines more difficult.

The last of the new breed would be John von Neumann, who was born in Budapest, Hungary, and was considered a mathematical genius already by the age of 10. Coming from a wealthy family, he had private tutors in addition to attending the proper schools. He attended university both in Budapest and in Zürich, finishing his doctorate in 1926 (from Budapest). As a result, he would arrive on the scene after the major discoveries in the new quantum mechanics, but would still provide a solid mathematical footing for the work. Following his doctorate, he went to Göttingen with a fellowship to work with Hilbert, but would spend plenty of time interacting with Born's young scholars. In 1928, he became a young faculty member in Berlin, but moved the next year to Hamburg briefly, before emigrating to Princeton University and the Institute for Advanced Study. His importance to quantum mechanics came during his period in Göttingen, where he made major advances in the mathematical foundations of the new field, which will be discussed in the next chapter. Throughout his career, he made major contributions to mathematics, including establishing game theory, quantum logic, mathematical economics, and to computers. His stored program, linear processing concepts are today known as the Von Neumann architecture. But, he also introduced the ideas of parallel processing, cellular automata, and neural networks. In addition, he introduced the Monte Carlo simulation method in studying neutron transport as part of the Manhattan Project, in which he was involved due to his expertise in the mathematics of explosions.

The accumulation of talent in Göttingen, coupled with the already existent expertise in mathematics, would prove crucial to the development of quantum theory. But, it would be an enigma to Bohr, as his ego would not accept the rise of a second center for quantum theory which threatened his dominance in Copenhagen. Yes, these young men came to Copenhagen to confer, but Göttingen would receive the credit. Would the guiding light still be Bohr, or would the mantel pass to Born. This would be decided in the next few years.

Von Neumann was not the only one to exit Germany during the coming period of problems. Max Born went to the United Kingdom and returned to Germany only after retiring. Wolfgang Pauli also moved to the United States, but later returned to Switzerland. On the other hand, Pascual Jordan became a leader in the new university movement under the new regime while Werner Heisenberg headed the German nuclear weapon program. Schrödinger would move to Ireland, but return to Austria after the war.

References

1. J. Dalton, Memoirs, *Lit. Philos. Soc. Manchester*, Ser. 2, **1**, 244 (1805).
2. R. Brown, *Philos. Mag.* **4**, 161 (1828).
3. A. Einstein, *Ann. Phys.* **322**, 549 (1905).
4. J. B. Perrin, *Les Atomes* (F. Alcan, Paris, 1920).
5. J. J. Thomson, *Proc. R. Inst. G.B. Irel.* **15**, 419 (30 April 1897).
6. J. J. Thomson, *Philos. Mag.*, Ser. 5, **44**, 269 (1897).
7. J. J. Thomson, *Philos. Mag.*, Ser. 6, **11**, 769 (1906).
8. H. Geiger and E. Marsden, *Proc. R. Soc. A* **82**, 495 (1909)
9. H. Geiger, *Proc. R. Soc. A* **83**, 492 (1910).
10. E. Rutherford, *Philos. Mag.*, Ser. 6, **21**, 669 (1911).
11. H. Nagaoka, *Philos. Mag.*, Ser. 6, **7**, 445 (1904).
12. A. J. Ångström, *Philos. Mag.*, Ser. 4, **24**, 1 (1862).
13. A. J. Ångström, *Recherches sur le spectre solaire* (W. Schultz, Uppsala, Sweden, 1868).
14. J. J. Balmer, *Basel Verh.* **7**, 548 (1885); **7**, 750 (1885); **11**, 448 (1897).
15. J. R. Rydberg, *Royal Swedish Vet.-Akad. Handl.* **23**(11), 1 (1889).
16. F. Paschen, *Ann. Phys.* **332**, 537 (1908).
17. T. Lyman, *Nature* **93**, 241 (1914).
18. L. Rosenfeld, Osiris **2**, 149 (1936); trans. in R. S. Cohen and J. J. Stachel, eds., *Selected Papers of Leon Rosenfeld* (Reidel, Dordrecht, 1979).
19. A. Pais, *Niels Bohr's Times, in Physics, Philosophy, and Polity* (Clarendon Press, Oxford, 1991), p. 15.
20. S. Jones, *The Quantum Ten* (Thomas Allen, Toronto, 2008).
21. G. Holton, *Thematic Origins of Scientific Thought* (Harvard Univ. Press, Cambridge, 1973), p. 130.
22. J. T. Cushing, *Quantum Mechanics* (Univ. Chicago Press, Chicago, 1994), pp. 101–102.
23. N. Bohr, *Philos. Mag.*, Ser. 6, **26**, 1 (1913).
24. A. Pais, *Niels Bohr's Times, in Physics, Philosophy, and Polity* (Clarendon Press, Oxford, 1991), p. 14.
25. G. Holton, *Thematic Origins of Scientific Thought* (Harvard Univ. Press, Cambridge, 1973), p. 104.
26. N. Bohr, *Philos. Mag.*, Ser. 6, **26**, 476 (1913).

27. H. G. J. Moseley, *Philos. Mag.*, Ser. 6, **26**, 1024 (1913).

28. A. van den Broek, *Phys. Z.* **14**, 32 (1913).

29. H. G. J. Moseley, *Philos. Mag.*, Ser. 6, **27**, 703 (1914).

30. A. Sommerfeld, *Ann. Phys.* **51**, 1 (1916).

31. P. Zeemann, *Nature* **55**, 347 (1897).

32. S. Jones, *The Quantum Ten* (Thomas Allen, Toronto, 2008), pp. 34–35.

33. S. Brush, *Graduate J.* **7**, 477 (1967).

34. P. Frank, *Between Physics and Philosophy* (Harvard Univ. Press, Cambridge, 1941), p. 117; cited in S. Jones, *The Quantum Ten* (Thomas Allen, Toronto, 2008), p. 174.

35. G. Holton, *Thematic Origins of Scientific Thought* (Harvard Univ. Press, Cambridge, 1973), p. 38.

36. P. Frank, *Modern Science and Its Philosophy* (Harvard Univ. Press, Cambridge, 1949), p. 115.

37. A. H. Compton, *Phys. Rev.*, Ser. 2, **14**, 20 (1919); **14**, 247 (1919).

38. http://www.nobelprize.org/nobel_prizes/physics/laureates/1927/compton-bio.html.

39. A. H. Compton, *Phys. Rev.*, Ser. 2, **21**, 483 (1923).

40. S. N. Bose, *Z. Phys.* **26**, 178 (1924); **26**, 384 (1924).

41. A. Einstein, *Ann. Phys.* **20**, 199 (1906).

42. A. Einstein and L. Hopf, *Ann. Phys.* **33**, 1096 (1910).

43. A. Einstein, *Phys. Z.* 18, 121 (1917); trans. in B. L. van der Waerden, *Sources of Quantum Mechanics* (Dover, Mineola NY, 1967).

44. L. J. Boya, *Int. J. Theor. Phys.* **42**, 2563 (2003); the citation there is credited to Pais, ref. 19 above.

45. J. C. Slater, *Nature* **113**, 307 (1924).

46. M. Jammer, *The Conceptual Development of Quantum Mechanics* (McGraw-Hill, New York, 1966), p. 182.

47. N. Bohr, H. A. Kramers, and J. C. Slater, *Philos. Mag.*, Ser. 6, **47**, 785 (1924).

48. M. Jammer, *The Conceptual Development of Quantum Mechanics* (McGraw-Hill, New York, 1966), p. 183.

49. Letter from Slater to van der Waerden, cited in B. L. van der Waerden, *Sources of Quantum Mechanics* (Dover, Mineola NY, 1967), p. 13.

50. AIP oral histories: https://www.aip.org/history-programs/niels-bohr-library/oral-histories/4892-1

51. L. J. Boya, *Int. J. Theor. Phys.* **42**, 2563 (2003); the citation there is credited to Pais, ref. 19 above.

52. M. Gell-Mann, in *The Nature of the Physical Universe: 1976 Nobel Conference*, ed. D. Huff and O. Prewett (Wiley-Interscience, New York, 1979), p. 29.

53. S. Jones, *The Quantum Ten* (Thomas Allen, Toronto, 2008), p. 103.

54. W. Bothe and H. Geiger, *Z. Phys.* **26**, 44 (1924).

55. W. Bothe and H. Geiger, *Naturwiss.* **13**, 440 (1925).

56. J. T. Cushing, *Quantum Mechanics* (Univ. Chicago Press, Chicago, 1994), p. 20.

57. P. McEvoy, *Niels Bohr: Reflections on Subject and Object* (MicroAnalytix, San Francisco, 2000), p. 40.

58. S. Jones, *The Quantum Ten* (Thomas Allen, Toronto, 2008), p. 147.

59. L. de Broglie, *Nature* **112**, 540 (1923).

60. J. T. Cushing, *Quantum Mechanics* (Univ. Chicago Press, Chicago, 1994), p. 126

61. L. de Broglie, Nobel lecture: http://www.nobelprize.org/nobel_prizes/physics/laureates/1929/broglie-lecture.html

62. S. Jones, *The Quantum Ten* (Thomas Allen, Toronto, 2008), p. 150.

63. L. de Broglie, *Philos. Mag.*, Ser. 6, **47**, 446 (1924).

64. C. Davisson and C. H. Kunsman, *Science* **54**, 522 (1921).

65. C. Davisson and C. H. Kunsman, *Phys. Rev.* **22**, 242 (1923).

66. C. Ramsauer, *Ann. Phys.* **64**, 513 (1921); **66**, 546 (1922).

67. W. Elsasser, *Naturwiss.* **13**, 711 (1925).

68. M. Jammer, *The Conceptual Development of Quantum Mechanics* (McGraw-Hill, New York, 1966), p. 251.

69. G. P. Thomson and A. Reid, *Nature* **119**, 890 (1927).

70. https://www.youtube.com/watch?v=jvOOP5-SMxk.

71. M. Born, *Ann. Phys.* **30**, 1 (1909).

72. M. Born, *Phys. Z.* **10**, 1031 (1909).

73. W. Pauli, *Phys. Z.* **20**, 457 (1919).

74. S. Jones, *The Quantum Ten* (Thomas Allen, Toronto, 2008), p. 129.

75. W. Pauli, "Relativitätstheorie," in *Enzyklopädie der mathematishcen Wissenschaften* (B. G. Teubner, Leipzig, 1921), trans. G. Field as *Theory of Relativity* (Dover, New York, 1981).

76. J. T. Cushing, *Quantum Mechanics* (Univ. Chicago Press, Chicago, 1994), pp. 108–109.

77. W. Pauli, *Naturwiss.* **12**, 741 (1924).

78. The American Institute of Physics history program: https://www.aip.org/history/exhibits/heisenberg/p06.htm.

79. S. Jones, *The Quantum Ten* (Thomas Allen, Toronto, 2008), p. 141.

80. H. A. Kramers and W. Heisenberg, *Z. Phys.* **31**, 681 (1925).

81. The American Institute of Physics history program: https://www.aip.org/history-programs/niels-bohr-library/oral-histories/4575-1

82. P. A. M. Dirac, *Proc. Cambridge Philos. Soc.* **22**, 432 (1925).

83. J. Gribbin, *Erwin Schrödinger and the Quantum Revolution* (Wiley, New York, 2013).

84. W. J. Moore, *Schrödinger: Life and Thought* (Cambridge Univ. Press, Cambridge, 1989).

85. E. Schrödinger, *Phys. Z.* **23**, 301 (1922).

Chapter 3

Arrival of the New Quantum Theory

As the second quarter of the twentieth century dawned, it seemed that physics was living in a state of anticipation. The questions had been raised and left unanswered. The duality of waves and particles had been shown to be the proper state of matter. Light waves were carried by photons, and electrons were ushered around by matter waves. But for the latter, no theory existed as yet. In addition, a new model of the atom appeared, largely based upon Rutherford's planetary model. More importantly, this new model had distinct energy levels, which would give optical transitions that closely matched those described by the Balmer and Rydberg formulae. But there were great differences between the planetary model and what was known about the atom's structure already at the time of Rutherford.

These differences created a crisis in physics which was deep and significant [1]. If the atom is to be stable, then it must be hard and unchanging. Yet, a true planetary system is quite accepting of small or large changes. For example, if two solar systems, each described by a planetary model, interact by colliding with one another, both will undergo great changes. To the contrary, the atomic system was held together by electromagnetic forces. The interaction of two atoms passing near one another did not greatly affect their structure

The Copenhagen Conspiracy
David Ferry
Copyright © 2019 Pan Stanford Publishing Pte. Ltd.
ISBN 978-981-4774-75-8 (Hardcover), 978-1-351-20723-2 (eBook)
www.panstanford.com

before or after the interaction. So, the physics was quite different. This was confirmed by the coming of the Bohr model of the atom. The atom would emit or absorb energy when an electron moved from one energy level to another, and these levels were proscribed by the energy quantization. The atom could not give off half a photon, as that was not allowed. But the planetary system was not quantized and could emit or absorb energy in incremental amounts without any large change. Then, it was known from chemistry that atoms would combine into molecules which often had quite specific shapes. This implies that the atoms themselves might have very specific shapes associated with the possible orbits of the electrons around the nucleus. So, the connection between these two systems was only figurative. In reality, one could draw no conclusions about one from considerations of the other.

The lack of a theory for the matter waves was considered to be a real problem. As remarked in the first chapter, Planck had emphasized this problem in his Nobel lecture. Hence, the situation was critical, and the theory was needed. This would change during 1925–6, when a plethora of papers surrounded two distinct new theories of quantization, one focused on the particles and one focused on the waves. In addition, there would be a third view involving both particles and waves. However, in the 2–3 years just prior to this, there were a great many new experiments and ideas that led to a wider understanding of how quantum physics had to evolve. Let us begin with these preliminaries. We have to emphasize that the Bohr model was not complete, and was not regarded with great love by many good physicists. Arnold Sommerfeld, although impressed with the results, did not regard the approach and mathematics as being acceptable. Hence, he was not sure about the model, and would do his own development a few years later. Ehrenfest, the leader of the physics group in Leiden, and a good friend of Bohr, was more critical [2]:

> Bohr's work on the quantum theory of the Balmer formula . . . has driven me to despair. If this is the way to reach the goal, I must give up doing physics.

As a result, many of the new experiments and calculations were done to check Bohr's model as well as to study properties which were not treated in the model.

Preliminaries

The motion of the electron around the nucleus, whether in Rutherford's planetary model or Bohr's model, is known as a central force problem, since there is an attractive force between the electron and the nucleus. This force is directed entirely along the direct line between the two quantities. The central force problem is only one form of the natural mechanics that arise from problems formulated with Hamilton's equations of motion or Lagrange's equations of motion. Lagrange formulated his version of mechanics in the late eighteenth century and Hamilton followed him to formulate his approach to mechanics in the first third of the nineteenth century. In both approaches, one generally expresses the equation of motion of the mechanical quantity as a second-order differential equation, in the full three dimensions of physical space. Of course, these equations are representations of Newton's equations of motion developed in the seventeenth century. This is all a result of the fact that our notions of classical mechanics have existed since that time [3]. By the middle of the eighteenth century, the generalized approach to solving Hamilton's second-order differential equations had become known as the Sturm–Liouville problem, after two French mathematical physicists. In electromagnetics, the equations of this type that are most commonly encountered are Laplace's equation and Poisson's equation, where Pierre-Simon Laplace and Siméon Denis Poisson also were French mathematical physicists who did their work in the early 1800s.

By "the full three dimensions of physical space," it is meant the particular coordinate system in three dimensional space that is chosen for the problem. For example, it is known that Laplace's equation is separable in 13 different coordinate systems, while Poisson's equation is separable in only 5 different coordinate systems; the general wave equation of Maxwell's electromagnetics is separable in 11 of these coordinate systems [4]. What we know from the general Sturm–Liouville problem is that the choice of coordinate system leads to the particular type of special functions that will be involved in the solutions [5]. But, quite generally, the separation of variables approach leads to the recognition that the solution will involve three parameters, loosely appropriate to the fact that the solution exists in three dimensions. Whether we call these three parameters separation parameters or eigenvalues doesn't really matter. In fact, the concept of eigenvalues comes from linear algebra rather than quantum mechanics, and came into the latter through the use of linear algebra and matrices, as discussed later. The important point is that the solutions in three dimensions involve three such eigenvalues. But Bohr had only two!

In developing his atomic model, Bohr had shown that the energy was quantized, and that this led to the angular momentum being quantized as well [6]. But he mentioned only one angular momentum, and this suggested that the model was a planar model, as first discussed by Nagaoka [7]. But Nagaoka also discussed a spherical model in which the orbits of the electrons would not be planar. This introduces another angle to the problem, and one has to worry about how this new angle affects the various energy levels. As we discussed above, in general the solution of the central force problem involves three parameters, but Bohr had only two. The easiest way to think of the problem is to place the nucleus at the origin of the spherical coordinate system, which describes the relative motion of the electron around the nucleus. In spherical coordinates, the three coordinates are the radius and two angles. One angle is the declination away from the z axis (in the usual rectangular coordinates), while the other angle describes motion around the z axis. In a sense, the former angle yields the latitude, while the latter angle yields the longitude for a sphere of constant radius.

As mentioned above, Arnold Sommerfeld felt that Bohr's atom could only be approximately correct in that it really had only one quantum condition, other than the use of Planck's model [8]. He began to investigate it closer. In 1916, Sommerfeld published his important extensions of the Bohr atomic model [9]. One of the most significant changes was to introduce rotation around the z axis as a quantum number. But this introduction allowed the electrons to have elliptical orbits. In this, there was another important point. This was that the energy levels were really given by

$$E \sim \frac{1}{n^2} , \qquad n = k + l , \qquad (3.1)$$

where k was the radial quantum number and l was a quantum number for the declination angle (away from the z axis). Another important observation from this result is that l cannot be larger than n. In fact, if $n = 1$, then $l = 0$. This means that the lowest energy level can have *no angular momentum*. Hence, Bohr was wrong in his estimates of the angular momentum. The quantization of the motion around the z axis leads to an angular momentum given by $mh/2\pi$, where m is an (positive or negative) integer whose magnitude is limited by l. Since this integer does not appear in the energy, it means that many values of m have the same energy, so that the concept of degeneracy arises from Sommerfeld's development. This motion of the electron around the z axis produces a z-directed magnetic moment for the atom, so that m is usually called the magnetic quantum number. Sommerfeld also realized that there is a maximum charge on the nucleus that will allow a stable electron orbit, and this maximum charge defines the fine structure constant

$$\alpha = \frac{1}{4\pi\varepsilon_0}\frac{e^2}{\hbar c}, \qquad \hbar = \frac{h}{2\pi}. \tag{3.2}$$

(In the second equation, we have introduced the *reduced* Planck's constant, which is used more commonly.) This constant is the ratio of the potential energy of the energy level at this maximum orbit radius d to the energy of the photon that would be emitted from this level, as

$$\alpha = \frac{e^2}{4\pi\varepsilon_0 d}\bigg/\frac{hc}{\lambda} \sim \frac{1}{137} \tag{3.3}$$

under the condition that $2\pi d = \lambda$. In a sense, this latter relation predates de Broglie's work (discussed in the previous chapter), which pointed out that the circumference of the orbit should be a multiple of the wavelength of the electron. Here, however, this wavelength is that of the photon, as the second expression is just Planck's law $E = hf$. But we cannot escape the conclusion that the expression is likely to have had an effect on de Broglie's thinking a few years later, and indeed that was the case.

Another important aspect is that the application of a magnetic field will break the degeneracy of the energy levels, and thus will lead to splitting of the spectral lines that are observed. It may have been discovered as early as 1891, when Michelson discovered that the Balmer series was not really composed of simple lines [10]. We first talked about the splitting of the spectral lines as due to spin, a point to which we will return. In reality, Zeeman is credited with this splitting in the presence of a magnetic field. The Zeeman effect refers to the splitting into doublets or triplets (or even more spectral lines). This effect arises because the magnetic field splits the symmetry of the system in which the atom is located. This splitting is commonly called the normal Zeeman effect, as it was discovered in experiments carried out by this Dutch physicist [11, 12], as we have discussed in the previous chapter. Later, Joseph Larmor showed that this splitting depended upon the charge-to-mass ratio of the electron [13]. Zeeman had trouble in fully resolving the "broadened line" into the doublets or triplets. His experiments were repeated by Thomas Preston through a herculean effort. He used a concave grating with a diameter of 21.5 feet, located in the Physical Laboratory of the Royal University of Ireland [14]. In his experiments, he discovered two effects. His first result was that a single spectral line would split into a doublet or triplet of closely spaced lines, and which he was able to see. In addition, however, he saw a doublet form even for [14] "the light being viewed across the lines of force." Moreover, he established that [14]

> such laws, therefore, as that the broadening of the spectral lines is proportional to the wave-length, or to the square of the wavelength are shown to be utterly untenable.

The latter doublet has come to be known as the anomalous Zeeman effect[a] and can be observed even when the light is not parallel to the magnetic field. It is this effect which was found to be due to spin, a point we return to below.

So, the Sommerfeld model went a long way toward cleaning up the deficiencies of Bohr's model. It explained a great many of the remaining questions about the optical spectra of the atom. Moreover, it was done with what could be called "proper mathematics" and arrived at the normal three quantum numbers. But it was still a semi-classical model. Ehrenfest was still grumpy, and wrote to

---

[a]There appears to be some differences in the literature as to the designation of *normal* and *anomalous* to the Zeeman effect. For example, the opposite choice is given in Ref. [8]. My choice is therefore somewhat arbitrary.

Sommerfeld to congratulate him on the new model in a roundabout manner [15]:

> Even though I consider it horrible that this success will help the preliminary, but still completely monstrous, Bohr model on to new triumphs, I nevertheless heartily wish physics at Munich further successes along this path.

In later years, Ehrenfest would soften and come to grips with the model, eventually coming to terms with its wide usage. Whether he really softened, or just wanted to retain his friendship with Niels Bohr, is an open question.[b]

Otto Stern received his doctorate from the University of Breslau in 1912, then moved to Prague to work with Einstein. He later did his habilitation at the University of Frankfurt, working part of this time with Max Born. In 1921, he became professor at the University of Rostock. However, he continued to work in Frankfurt, primarily with Walter Gerlach. In 1921, Stern had proposed trying to measure the spatial quantization that appeared in the Bohr and Sommerfeld models, and thought that this could be done with the use of a non-uniform magnetic field. Gerlach had finished his habilitation in 1916, during the war, at the University of Tübingen, then moved to Frankfurt in 1920. In their experiment, they created a very non-uniform magnetic field, and sent a collimated beam of Ag atoms through the magnetic field. In a normal system with a magnetic moment, as that supposedly existing the Bohr and Sommerfeld atoms, a magnetic field will cause the magnetic moment to precess, much like a top where the axis of rotation moves around the vertical direction as it wobbles. In an inhomogeneous magnetic field, however, the precessional torque that is applied to one end of the object is different from that of the other end and the object will be deflected in the direction of the magnetic field. In the classical models, the actual direction of the magnetic moment is random from atom to atom and the observed impact of the deflected atoms should be a vertical line created as they impact a screen. This line has a finite

---

[b]Ehrenfest was certainly a friend of Bohr, and thus he probably did not want to submit to the kind of arguing for which Bohr was noted. He is thought to have suffered from depression, and in 1933, he took his own life and that of his disabled son.

length, since the magnetic moment has a maximum value. What they found, however, was only two spots, centered upon the axis of the collimated beam [16]. As Rammer says [17], "taking exposures eight hours or more, Stern and Gerlach established unequivocally that the atomic beam split in the magnetic field into precisely two beamlets." The remarkable thing is that no atoms were discovered which did not deflect. This meant that the magnetic moment being measured did not depend upon the principal axis (e.g., the z axis of each atom) or orientation of the individual atoms. While the results were considered to be a confirmation of the Sommerfeld atom, this went beyond classical physics.

In 1926–7, the Stern–Gerlach experiment was repeated, at the University of Illinois, by T. E. Phipps and J. B. Taylor using hydrogen atoms. In this experiment, the hydrogen atoms were prepared in a manner such that they were relaxed into their ground state [18, 19]. This would assure that only a single electron was involved, so that any questions due to the multiple electrons in the Ag atoms could be put aside. By this time, these latter authors were able to interpret their results within the new quantum mechanics and ideas of the electron spin, which we shall come to shortly.

Wolfgang Pauli finished his doctorate in Munich under Arnold Sommerfeld in 1921. As we have already noted, he then spent a year in Göttingen with Max Born and then a year in Copenhagen with Niels Bohr. In 1923, he took up his own faculty position in Hamburg. He was known as an arrogant individual with no real respect for authority, some of which was discussed in the previous chapter. By late 1924, Sommerfeld's set of three quantum numbers and the duality of results from the Stern–Gerlach experiment were known throughout the community. Pauli was in Hamburg at the time when he realized that the two were connected. He then proposed that the electron in the atom was characterized by four quantum numbers: Sommerfeld's three and a new one characterizing the nature of the results from the Stern–Gerlach experiment. He referred to the latter as a kind of two-handedness, e.g., either right-handed or left-handed. Hence, this latter quantum number could have only two values. This led him to consider what became known as an exclusion principle. Rosenfeld has pointed out [20]:

> The exclusion principle ... gives rise to an additional interaction ... called exchange interaction to remind us that it has its origin in the identity of the particles, and not at all because it could be attributed to any actual permutation of their locations.
>
> ... it illustrates the fact that the matter waves can transmit force just as well as the electromagnetic waves."

Heisenberg would say later [21]: "I know that Pauli was very proud of the exclusion principle paper just on account of the fact that he could disprove some of the things which Bohr had claimed." The reason for this was [22]

> because it was incompatible with Bohr's orbital model. Bohr and Pauli joked, with no small measure of frustration, about the "nonsense" and "swindle" of the current theory of quantum physics, with its untenable mixture of quantum and classical physics to manipulate a result so as to match experimental evidence.

Already, Edmund Stoner [23], a British physicist working at the Cavendish Laboratory in Cambridge, had discussed how electrons could be distributed within an atom, corresponding to the Sommerfeld model. Stoner used a notation that had been put forward by Landé [24], and then proposed that each quantum state could hold only two electrons. This would lead to the well-known 2, 8, 18, ... sequence of occupancy in the shells of that atoms. Already, this requires that each eigenstate must have four quantum numbers, only one electron can have each distinct value of these. But Pauli [25] is credited with the realization that the shell structure of atoms would have the natural explanation if there were four quantum numbers with only a single electron in each possible state. Here, the fourth quantum number is his handedness. So, should Stoner or Pauli actually be credited with the exclusion principle? Pauli, with his connections to the Copenhagen–Göttingen pipeline, certainly has received the majority of the credit, and eventually the Nobel prize in 1945 for the exclusion principle. But Stoner never discusses a reason for the factor of 2 multiplicity, although he does talk about limits on the number of electrons per set of inner quantum numbers ("inner" here refers to the spatial eigenvalues). The idea that the factor of 2 denotes an additional quantum number thus seems to be that of Pauli alone.

The idea of arranging the elements into a "periodic table" was pushed by Bohr in the early twenties. The idea was, of course, much older. The concept appears to have begun with Lavoisier. But the British chemist John Newlands is thought to have published the first such table in which he ordered the elements by their atomic weight [26]. The idea of organizing a table in order of the atomic weights is slightly older, as a French geologist, de Chancourtois, is thought to have suggested the idea but never published his table. Newlands went on to discuss his table as a law of octaves by which the critical number of 8 would create some periodicity. But it would be the Russian chemist Dmitri Mendeleev who would be recognized for the table [27], although he retained the ordering based upon the atomic weight, which continued the idea that this had something to do with the chemical properties. These early tables, with their reliance on the atomic weight, had a number of problems that arose from this assumption. We met Henry Moseley earlier, where we discussed his discovery of the fact that there was a relation between the X-ray wavelength of atomic emission and the total nuclear charge, rather than the atomic weight [28]. Moseley was then able to sort the elements based upon their atomic number (nuclear charge) rather than the atomic weight, which gave a new table that agreed much better with the chemical properties of the elements. Bohr, however, recognized that these periodic tables were empirical, and thought that they should arise in a better form from his atomic theory [29]. Here, Bohr relied upon the noble gases He, Ne, and Ar to serve as the ends of a row in the table. What was important is that these elements have 2, 8, 18 electrons in their full set of shells. Bohr had first addressed the idea of assembling the atoms into an atomic table in 1913, in his second key paper [30]. This was just before Moseley's paper appeared, and before Sommerfeld's corrections to the atomic model were published. Thus, he was still thinking of planar rings of electrons, and proposed "the numbers of electrons on inner rings will only be 2, 4, 8, ..." With his later paper [29], he apparently accepted the advances made by Sommerfeld and

Moseley, as he argued for the 2, 8, 18, . . . configuration as being the natural approach. But, as Heisenberg later recalled [31], "Bohr didn't have a real explanation of the closed shells, and he only could get pictures which sounded reasonable, but he didn't really understand. He couldn't really understand." Bohr then discussed the inert gases, and showed how their electronic arrangement fit into this system. Nevertheless, his paper is usually not mentioned in any history of the periodic table, perhaps for the reasons Heisenberg mentioned. While the possible number of electrons per shell proposed by Bohr would be used by Edmund Stoner, it also brought the physicists worrying about the electronic structure into the game with the chemists, who worried more about the chemical reaction properties of the elements. The work of Stoner and Pauli now laid the basis for developing the full periodic table, although of course, later work would address curious problems that arose from partially empty inner shells. No matter the success of Stoner and Pauli, there was still no understanding of what the handedness really was.

Ralph Kronig was a young German-American, as we have already mentioned in the previous chapter. While a student at Columbia, he had met Paul Ehrenfest during a visit of the latter to the university. This encounter led to Kronig going to Europe while finishing his thesis and serving as a "traveling fellow" of Columbia. Thus, in late 1924, he was working with Landé in Tübingen. It was in January of 1925, that Pauli visited and gave a lecture on his ideas on atomic structure. Prior to this, Pauli had communicated his ideas in a letter to Landé discussing his four quantum numbers and the exclusion principal [32]. The letter was shared with Kronig, who began to believe that the fourth quantum number could represent an intrinsic angular momentum arising from a spinning motion of the electron, and he carried out preliminary calculations to support the idea. However, in various discussions with Pauli, Kramers, and Heisenberg, his ideas were generally rejected. That is, they were somewhat rudely rejected by Pauli, when Kronig raised the question following the lecture in January. The alleged reason for this was that such a spin would cause the rotational speed of the electron to be larger than the speed of light, or so it was supposed. As a result, Kronig never published his ideas. However, it appears that the idea was actually somewhat older. A. L. Parson, a British chemist, had speculated that the atom would have a magnetic moment which would correspond to an elementary particle spinning about its own axis, with a velocity at the circumference equal to the velocity of light [33]. While he did not mention the electron specifically, the similarity is sufficiently close to suggest that the arguments of the doubters were off base.

But ideas often are stimulated at a variety of places, and Pauli's paper did just this. Samuel Goudsmit was a doctoral student of Paul Ehrenfest in Leiden. George Uhlenbeck had been a student of Ehrenfest earlier, but had gone to Rome to tutor the son of the Dutch ambassador. There, he had met Enrico Fermi. He had just returned to Leiden in 1925 to become Ehrenfest's assistant. Goudsmit and Uhlenbeck were tasked with the problem to find what was new in physics, and this led to Pauli's paper on the exclusion principle. They proceeded to do similar calculations to Kronig (although it is generally believed they were unaware of Kronig's results), which led them to believe that the additional magnetic moment was the electron spinning on its own axis. They were not put off by obtaining velocities above the speed of light, which arose only if they considered the electron to be a true point particle. Earlier, Bohr had proposed that the two handedness could be explained by an additional nonmechanical stress that would exist within the atom [34]. Consequently, Uhlenbeck and Goudsmit explicitly addressed this in their first paper, where they proposed to replace the nonmechanical stress with the electron spin [35]. However, the theoretical predictions for the precession of the angular momentum remained too high by a factor of 2 [36]. The following year, Llewellyn Thomas, a young British physicist studying at Cambridge, pointed out that there should be a relativistic correction that would add a factor of 1/2, and this problem was solved [37] (we know more of Thomas for his contribution to what we call the Thomas–Fermi model, which led to density functional theory). But there remained the problem with the velocity. And, how did the electron spin couple to the other (orbital) angular momentum in the atom? The latter question was answered by Einstein, who pointed out that, in relativity theory, a moving object in an electric field would generate an effective magnetic field [36]. This magnetic field would provide the needed coupling. The definitive answer to the velocity problem would come later, and we leave it to a later point in this book, where we discuss Dirac's relativistic formulation of quantum mechanics. However, we have to recall that Compton had already predicted that the size of the electron was perhaps two orders of magnitude larger than the classically expected size [38]. Arriving at a velocity above the speed of light was based upon the smaller classical size, and one would need to ask what the peripheral velocity would be if the size were larger than expected. Nevertheless, it was clear that the handedness of Pauli was related to an angular momentum, which was explainable by the spin of the electron on its own axis. Yet, Pauli was reluctant to accept this explanation, still bothered by the velocity. As Heisenberg later recalled [39],

> Pauli was perhaps the last of all the important physicists to be convinced.... Pauli perhaps was not so glad about the electronic spin just because it was too classical again. He preferred his degree of freedom which one could consider as something only quantum theoretical.

Heisenberg's Kinematics

In the Fall of 1924, Werner Heisenberg was still in Copenhagen with his Rockefeller Foundation fellowship, and doing battle with Kramers, as discussed in the previous chapter. But his mind was still working. It was becoming clear already at this time that the old quantum theory was not working for a significant number of experiments, and that something new was needed. Not coincidentally, these various possible approaches were going through Heisenberg's mind, as well as the minds of the other players in the field. But Heisenberg's were perhaps a little further afield from the classical ideas, and led to many arguments and discussions with Bohr. As Heisenberg himself later remarked [40]: "The contradictions between this Bohr–Sommerfeld quantum theory and classical theory were so terrible that one had to find excuses." Heisenberg would have many argumentative discussions with Bohr during his stay, but he was quite headstrong and would continue until he felt he had achieved some agreement. As he put it [40],

> after eight or nine o'clock in the evening, Bohr, all of a sudden, would come up to my room and say, "Heisenberg, what do you think about this problem?" And then he would start talking and talking and quite frequently we went on till, twelve or one o'clock at night. Or, sometimes, he would call me to his flat near the institute—at that time he stayed at this flat there—and finally at one o'clock at night we would feel that we were tired and would take a glass of port wine and then would go to bed.
>
> I remember that I tried to explain this to Bohr and Bohr first was impressed and he said, "Well, that looks quite all right." . . . But then after I think one day, I was called to Bohr's room, and Bohr and Kramers were there, and they now tried to explain very seriously that it was all wrong, that this idea didn't work. And I was completely shocked. I got quite furious because there I thought I had something real, and now they tried to explain it away. So we had quite a heated discussion, but at the end I think I came out with a slight victory because in some way I had succeeded in convincing at least Kramers. It was easier to convince Kramers than to convince Bohr. Well, the end of this discussion was that they said, "Well, we must think about it again, and next day or so we all agreed that this was a possibility.". . . I had, for the first time, the feeling that now I had been able to convince Bohr of something about which we had disagreed.
>
> I have really, in this whole period, been in real disagreement with Bohr; and the most serious disagreement was [later] at the time of the Uncertainty Relations. But this was the first time that I was really disagreeing with him and where I really got very angry about Bohr. You know sometimes, in some way, I was quite offended.

In many ways, the early arguments came to a head with the paper that Kramers eventually published with Heisenberg (mentioned in the previous chapter), although they would continue whenever Heisenberg was in Copenhagen. As Heisenberg recalls it [40],

> I don't know whether by Kramers himself or by somebody else— it was suggested perhaps it was just good that Kramers should

publish it alone because most of the work was due to him. I was a bit hurt by this idea because I felt that by insisting on this business with the one term I had made an actual contribution. But I simply said, "Well, I leave that entirely to Bohr, as I had left it always to Sommerfeld. Bohr shall decide. I don't care which way his decision goes." Bohr then probably thought, "Well, after all, the young man has contributed a bit, why not put his name also on it. Everybody knows that it's really Kramers' theory because Kramers had written the first note." So this must have been Bohr's idea more or less. I think that was quite right. So that was the history of the paper. I also believe that Kramers first had thought he would work out this whole thing alone but then the Raman lines came in as a kind of surprise which had to be worked into it. Since I always discussed these things with Bohr and Kramers, it was natural that I would come into the game. Then I had very early insisted on a special point which then turned out to be correct and in this way I came into the paper.

In the early spring of 1925, the Kramers–Heisenberg paper appeared and Heisenberg returned to Göttingen for his duties there. Nevertheless, the arguments in Copenhagen seemed to gnaw on Bohr, and he apparently was concerned over Heisenberg going off into directions which Bohr felt were "improper." He even asked Pauli, now in Hamburg, to intercede to see if he could lead Heisenberg back into the way of righteousness.

Born had been planning to spend the previous fall in visiting the United States, but gave up the trip to stay and work. Apparently, winter and spring were terrible allergy seasons in 1924–5, because he was nearly incapacitated by his severe asthma attacks. Then, when Heisenberg returned to Göttingen, his own health began to deteriorate significantly for the same reasons. Still also suffering from the stress of dealing with Bohr in Copenhagen, he decided to try to get away to both think and recover from his allergies. He chose to go to Heligoland, a small pair of islands in the North Sea close by Germany. The islands were basically free of any pollen and washed over by sea breezes which helped keep the air cleaner.

Heligoland is an unusual outcropping of rock in the North Sea. And, apparently, the nature of the rock itself is unique in this region. Originally a single island, it had been split apart by a storm surge in the early 1700s. At the time, the island apparently belonged to Denmark. The islands lie about 30 miles off the mainland, and the main port point of access is from Coxhaven, where the Elbe flows into the North Sea, which is just over 40 miles away. Although isolated, it was an important site, as it was the harbor in which ships would pick up their pilots for transit into the ports of the Hanseatic League. After the Napoleanic Wars, the islands passed into British hands. Surprisingly, it seems that the words to the German national anthem were written by August Heinrich Hoffman von Hallersleben while visiting these British held islands (the music is from Haydn). So, it seems that this was a good site for creativity as well as rest and relaxation. It was while under British rule, that the islands became a popular tourist attraction for those who could afford the costs of travel to them. Heligoland returned to German rule in 1890 when an accommodation was made between Germany and Britain. Even now, the population is not large, and there are only a few tourist hotels and apartments. There were probably fewer of these in 1925. So, many guests stayed in the equivalent of a bed and breakfast (and likely got all their meals at the site). The islands would be perfect for Heisenberg.

The first task would be relieving his allergies, and this took a few days, but this also gave him time to think, and it was here that he formulated his new view of quantum mechanics. Today we think of this as the new quantum mechanics, but he thought of it as just a new advance in the single flow of quantum mechanics that had begun with Planck. Einstein, and Bohr. It is not clear whether he viewed the new ideas as a break from the old views. He had come to Heligoland in early May, and by June he was both recovered from the allergies and ready to discuss his new approach. He first discussed the paper with Pauli. He knew Pauli already from his student days in Munich and considered him almost as an older brother. Thus, he wanted to run his ideas past Pauli to make sure that they were not total rubbish. Pauli was rather encouraging, so Heisenberg gave the paper to Born in early July to read, but Born found it mystifying and took some time to work his way through it, meeting Pauli again on a train [41]: "To Born, the mathematics in Heisenberg's paper obviously needed work, so he asked Pauli about collaborating with him on the changes. Pauli refused, responding in his typically blunt fashion, 'I know you are fond of tedious and complicated formalism. You are only going to spoil Heisenberg's physical ideas by your futile mathematics.'" Back in Göttingen a few days later, Born went over Heisenberg's paper again and then sent it off for publication. Heisenberg would later summarize this continuing disagreement between Born and Pauli as follows [42]:

> In Göttingen Born liked complicated, elaborate mathematical formalism. This was an idea which Pauli disliked definitely because he felt that the real problems are not those of mathematics—the real problems are those of physics.

This would lead to a somewhat angrier response of Heisenberg later.

The paper was received by the journal on July 29, and appeared in print in the December issue of *Zeitschrift für Physik* [43]. In the very first sentence, Heisenberg expressed the philosophical basis for the paper:

> The present paper seeks to establish a basis for theoretical quantum mechanics founded exclusively upon relationships between quantities which in principle are observable.

It is clear from this first line that he had been thoroughly imbued with Bohr's view of reality, giving the importance of the experiment priority. As remarked by Jammer [44],

> if Heisenberg's conception of the very possibility of rejecting the description of atomic systems in terms of classical physics may be traced back . . . via Bohr, . . . his choice of the particular nature of the new concepts by which he replaced the classical ones goes back to . . . the positivist or logical empiricism of the early twenties.

Indeed, Heisenberg's paper spends considerable time comparing the classical approach and formulas with new quantum mechanical formulas, but there were some conditions on the new derivations. First was the requirement that, for large quantum numbers, the results should correspond to equivalent processes found in classical mechanics. That is, the role of quantization would likely be smeared out in states with large quantum numbers, and only in the true lower quantum numbers of the small atoms, like hydrogen, would true quantum mechanics deviate from the classical results. Moreover, he justifies the new approach on the grounds [43]:

> The Einstein–Bohr frequency condition (which is valid in all cases) already represents such a complete departure from classical mechanics, or rather (from the viewpoint of wave theory) from the kinematics underlying this mechanics, that even for the simplest quantum-theoretical problems the validity of classical mechanics just cannot be maintained.

This point is important, as Heisenberg's new theory would be a kinematical[c] theory rather than just a mechanical theory. The phrase in the second set of parentheses is interesting, as it appears at this time, Heisenberg was quite accepting of the wave idea in his own theory; he would later never accept the wave theory for quantum mechanics that Schrödinger would obtain beginning with de Broglie's wave ideas. The concepts of the paper address how to replace the classical approaches with his new quantum approaches. For example, if one takes a simple linear (second-order) differential equation, so that the motion is periodic in time (here, the connection to the oscillators of the previous theories is incorporated), then one can develop the solutions in terms of a Fourier series as

$$x(t) = \sum_n a_n e^{in\omega t}, \tag{3.4}$$

where x is the variable from the differential equation, n is an integer that runs over all possible values and ω is the fundamental frequency of the oscillator. This expression should now be replaced in the new quantum theory by

$$x(t) = \sum_n a_n e^{i\omega_n t}, \tag{3.5}$$

where now, the ω_n correspond to the energy values of the states in the system. This would lead directly to the result, in accordance with Bohr, that the frequency of the transition was given by

$$hf(n, n - m) = E(n) - E(n - m), \tag{3.6}$$

---

[c]Kinematics is that branch of classical mechanics which involves motion. Here, this means that the momentum would be an important part of the theory.

where n and m are integers describing particular energy levels (we are using Heisenberg's notation here), and E is the energy of these levels. Thus, he would then introduce the transition probabilities as quantities that varied as

$$a(n, n - m)e^{i(\omega_n - \omega_{n-m})t}. \tag{3.7}$$

If the system in state n was the ground state (lowest energy level), then it stood to mean that the transition rate $a(n, n - m) = 0$. Another crucial point is that the relationships between frequencies in Eq. 3.7 and their connection to energies in Eq. 3.6 meant that the energy was quantized into discrete values. Hence, the system was required to become quantized by the form of the new mathematics, but, once again, the actual energy levels for the hydrogen atom would be changed, and Eq. 3.1 became

$$E_n \sim -\frac{1}{n^2}, \qquad n = k + l + 1, \tag{3.8}$$

where n is the principal quantum number, k is the radial quantum number and l is the angular momentum quantum number. So, the energy has progressed from Bohr's simple version, to Sommerfeld's more common classical version, to the quantum mechanics version.

From the time of Planck, the properties of atoms and solids were discussed in terms of oscillators from which radiation was emitted and which, in turn, was absorbed. In Heisenberg's treatment, with his new formulation, he found that the energy levels of the oscillators was not given by $n\hbar\omega$, as Bohr had asserted, but by $(n + 1/2)\hbar\omega$. Since $n = 0$ is allowed in this new expression, it seems that the available classical states are symmetrically compressed into a quantum state; e.g., states from both above and below a quantum state are quantized into the particular energy level. Hence, each quantum level has the same statistical weight. Now, Heisenberg could share Pauli's joy over having proved Bohr wrong in some of his assertions. Heisenberg was able to show that the dispersion of the optical emission and absorption in atoms was related to the square magnitude of the transition probability (or amplitude), but that the phase was, in principle, still measurable. In the end, he recovered the same energy levels for the hydrogen atom that Bohr had found, even with the correction to the oscillator energies. Yet, in this approach he introduced another symbolic problem. In making the transition that was induced by either the emission or absorption of radiation, the electron in the atom would have to "jump" from one energy level to the other. There was still no chance in the theory for a gradual transition in this process. The emission or absorption of a light corpuscle required this jump to occur.

The irony in Heisenberg's paper was the lack of rigorous mathematics. Heisenberg had demanded a high level of mathematics in his earlier work with Kramers. But, for his own paper, he made no such demands. Perhaps it was due to the fact that he was not familiar with the mathematics he demanded for the paper. It is generally believed that he had no acquaintance with matrix mathematics, and this could explain his reluctance. But he worked for Born, and yet he did not approach Born to help with the mathematics. Even when Born mentioned to Pauli, his desire to better develop the mathematics, he was rudely rebuked by the latter. We will see later that even Pauli's enthusiasm for the paper would dissipate within a few weeks.

What was the impact of the Heisenberg paper? Heisenberg had taken a very different step forward. As McEvoy says [45], "he labeled statements employing quantities which could not be measured according to quantum theory as 'having no sense'.... He maintained that in order for the phrase 'position of an electron' to have meaning ... [requires] experiments by which the position could be measured." Of course, some felt them to be revolutionary. But there were questioning comments as well. In Italy, Enrico Fermi wrote to his colleague Enrico Persico about what he called [46]

> the formal results in the zoology of spectroscopic terms achieved by Heisenberg. For my taste, they have begun to exaggerate their tendency to give up understanding things.

And others would later describe the process [47]:

> Various mathematical formalisms were devised which simply "described" atomic states and transitions, but the same arbitrary avoidance of detailed processes ... of the actual process of the transition, were inherent in all of these formulations.

Along this line, Paul Dirac would comment [48]

> with the question "Can an equation of motion be used?" Although
> agreeing that Heisenberg's 1925 view can be a good guiding
> principle (e.g., that only observable quantities should be used in
> formulating a physical theory), he felt it nevertheless unlikely that
> the analytic S-matrix description would be the final answer. . . .
> This . . . led Dirac to say "a theory that has some mathematical
> beauty is more likely to be correct than an ugly one that gives a
> detailed fit to some experiments."

Heisenberg had clearly taken quantum mechanics into a new
direction.

Born and Jordan

In 1925, Born was trying to recruit a new assistant.[d] He asked Pauli,
who refused, especially as he already had his own faculty position.
As Jammer puts it [44],

> Born, while traveling by train to Hanover, told a colleague of
> his . . . about the fast progress in his work but also mentioned the
> peculiar difficulties involved . . . with matrices. It was . . . almost
> an act of providence that [Pascual] Jordan . . . shared the same
> compartment [and] overheard this piece of conversation . . . [and]
> in Hanover Jordan introduced himself.

The chance encounter leading to a fruitful collaboration is almost
pure Hollywood. Born himself discusses such an encounter with
Jordan [49] and seems to indicate that it occurred after he (Born)
had received Heisenberg's paper. But this cannot be, for if we

---

[d]It is probably of interest at this point to describe "assistant." In the Germanic
countries, it is still the case that the "assistant" is a position beyond that of a postdoc,
and probably closer to what we would call "Research Assistant Professor." In this
position, he conducts research while sometimes also aiding the professor at lectures.
The research is usually in aid of his habilitation. Even after achieving this, he may
remain an assistant, corresponding perhaps to "Research Associate Professor," until
acquiring a faculty position of his own.

interpret the matrix comment to refer to Heisenberg's paper, it must combine at least two train rides. It is true that Born met Pauli on July 19, while traveling from Göttingen to Hanover, and discussed Heisenberg's paper [50], which he had only recently been given. It was here that Pauli made the comment, cited above, concerning Born's preference for mathematics, but he would not have met Jordan on this train for the first time. Jordan was working with Born much earlier. They had, in fact, a coauthored paper, which was published in December [51], but which had been received by the editor June 11, 1925. And Born himself talks about submitting this paper in his recollection with this date [49]. In addition, Born mentions working with Jordan in a letter to Einstein sent July 15 [52], and commenting on the paper that had been submitted already. Clearly, Jordan was working with Born prior to Heisenberg's return from Heligoland. Born may have met Jordan on a train to Hanover prior to this, as Jordan's family lived in Hanover. Moreover, the main route for trains between Göttingen and Copenhagen passes through both Hanover and Hamburg, but any discussion by Born on this earlier train could not have involved Heisenberg's paper. In fact, Born also mentions the Heisenberg manuscript in his letter to Einstein [52], and annotates the letter with the comments "Heisenberg gave me his manuscript on the 11th or 12th of July, asking me to decide whether it should be published and whether I had some use for it, as he was unable to get any further. Although I did not read it straight away, because I was tired, I had certainly read it before I wrote to Einstein on July 15th."

Jordan had become Born's assistant, likely well before June, and this led to fruitful work between them. It was after the train ride with Pauli, that the next day, July 20, Born turned to Jordan to begin the mathematical efforts [50]. It was clear that, although he allowed Heisenberg to submit his paper without what Born considered important changes, Born himself would work to make sure that these changes appeared in the literature as soon as practicable. Jordan provided him with the aid he needed to get this accomplished.

Within a few days of beginning to work on the task, Jordan brought to Born the first major result of the fact that, in general, matrices do not commute. That is, if we take two (non-diagonal) matrices A and B, the products AB and BA usually give different results. Jordan had approached the problem in a manner to first assure that Heisenberg's notation was consistent with matrix operations, and also consistent with Hamilton's canonical equations of motion, which would lead to classical behavior in the right limits. As a result, he achieved the important commutator relation

$$xp - px = i\hbar, \tag{3.9}$$

in which x and p are assumed to be operators rather than simple parameters. We have written the equation here as a scalar for simplicity, but it is a diagonal element, so that only the same components of each of the two operators satisfy this equation. He was able to directly achieve the proof of this formula as well as show that energy was conserved in Heisenberg's formulation. He then continued to develop the theory and was able to justify Heisenberg's suggestion that the transition probability would be given by the square of the magnitude of the matrix elements connecting the two states. An additional and very important result was that there was an induced time variation on operators, that were not explicitly functions of time. This arose through the above commutator relation as

$$\frac{dO}{dt} = \frac{1}{\hbar}(OH - HO), \tag{3.10}$$

where H is the total energy Hamiltonian function, which is quite similar in both classical and quantum mechanics, and O is any operator or linear operator expression (not itself an explicit function of time). These last two equations are perhaps the heart of the operator and matrix approach to quantum mechanics. If O and H are defined by the set of states in the system, then the products in Eq. 3.10 are matrix products.

In classical mechanics, there is an equivalent formulation, which is called the Poisson brackets [53]. In the classical world, Eq. 3.10 would be written as

$$\frac{dO}{dt} \sum_k \left(\frac{\partial O}{\partial q_k} \frac{\partial H}{\partial p_k} - \frac{\partial O}{\partial p_k} \frac{\partial H}{\partial q_k} \right). \tag{3.11}$$

Now, for the harmonic oscillator, in which the potential energy is quadratic in q (position), a connection is made to both Eqs.

3.5 and 3.6 if the conjugate variables in Eq. 3.5 are recognized as being differential operators. As such, it becomes quite natural that the ordering of the operators is important for the result. Just as in matrices, reversing the order gives a different result. Then, the matrices associated with these operators become connections between the different states in the system. In the quantum world, these operators can become complex quantities, which appears in Eq. 3.6 via $i = \sqrt{-1}$, and the existence of the quantization itself appears via Planck's reduced constant.

By early September, Jordan had completed the first two chapters of the resulting paper. In order to keep Heisenberg informed of the progress, Jordan sent these to him in Munich. The later chapters served to fill out the full mathematical treatment and to address a few examples. Moving quickly, Jordan and Born were able to complete their paper rapidly. Their paper [54] was received by the journal on September 27, just two months after the Heisenberg paper. It was also published in December 1925. While Born worked with Jordan on the mathematical paper, he still felt that he needed to get Heisenberg back into the loop. Thus, he put the two, Heisenberg and Jordan, together to work on developing the theory further, primarily as Jordan was an expert in the mathematics of matrices. Hence, this would aid in Heisenberg's learning about this methodology. In the middle of all of this, Born took family holiday and went to Switzerland. At the same time, Jordan went to visit his family in Hanover. For his part, Heisenberg would first visit his family in Munich, then go to Cambridge, to deliver his invited lecture, and finally return to Copenhagen, in order to finish his fellowship for study there. What made matters worse was that Born became physically ill while on his holiday, and was unable to contribute much to the scientific effort. So, Jordan was left with the lion's share of the effort in perfecting the derivations and writing the mathematical Born–Jordan paper. The collaboration with Heisenberg would depend upon long distance communication, especially after the late September submission of the Born–Jordan paper. This new effort would take them until the end of the year.

Born and Jordan provided the mathematical basis. More importantly, by providing the solid mathematical basis, Born had insured that Heisenberg's theory would be more acceptable to the community. Without this, the world may have passed him by. Now, however, Pauli's early encouragement had turned to a very dismissive attitude, as he didn't particularly like the Born and Jordan mathematics, even though it clearly supported Heisenberg's theory. In fact, Pauli's complaints finally made Heisenberg snap back at him [55]:

> Your endless griping about Copenhagen and Göttingen is an utter disgrace. Surely you will allow that we are not deliberately trying to ruin physics. If you're complaining that we're such big jackasses because we haven't come up with anything physically new, maybe you're right. But then you're just as much of a jackass, since you haven't achieved anything either.

It was an utter put-down for Pauli, since it referred to his own view of his successes or lack thereof, and Einstein certainly didn't like the deferral to only what could be measured. Heisenberg and Einstein would meet the following year and discuss this point. As Heisenberg recalled the discussion [56],

> you come very soon to the . . . view, "Well, why not say that all the things which should be handled in theory are just those things which we also can hope to observe somehow." That was perhaps wider spread in Gottingen than in Munich on account of this interest in relativity, and the point in relativity was emphasized quite strongly by Minkowski and such people when they gave lectures on it. I remember that when I first saw Einstein I had a talk with him about this. Einstein just explained to me that he did not agree with this point of view.... After some colloquium in Berlin, I went to Einstein's house; he said that he wanted to talk to me about quantum mechanics. I told him that this idea of observable quantities was actually taken from his relativity. Then he said, "That may be so, but still it's the wrong principle in philosophy." And he explained that it is the theory finally which decides what can be observed and what cannot, and, therefore, one cannot, before the theory, know what is observable and what not. On the other hand, of course, he agreed that as a heuristic principle it was extremely important. It was a way of finding what one should probably put into a theory.

Another interesting point about these new developments was that Bohr had been kept in the dark. He did not know about these new developments because Heisenberg had developed a little caution after Pauli's criticism of his new ideas. Hence, Bohr did not learn about these new approaches, and the new quantum theory, until Heisenberg arrived back in Copenhagen in mid-September [41]. At this time, there was little he could do to affect the new papers, so he decided to be enthusiastic in his support. However, he would later claim [57], "When Heisenberg came here, and also Dirac, they did wonderful things, but every word in any of their papers was obvious to me. Not that I had made it myself, but it was obvious that this was what we were waiting for."

The English Student

In the early Fall, Heisenberg was invited to Cambridge to give a lecture (before his paper was known to others than the Göttingen crowd), and then returned to Copenhagen to finish his fellowship. Nevertheless, Born had put him in touch with Jordan and they could correspond while developing the new quantum mechanics further, but Born was spreading the word. In the late summer, he had communicated with Einstein, and mentioned the Heisenberg manuscript. But, more importantly, he apparently sent a copy to his friend Fowler at Cambridge. As we learned earlier, Paul Dirac was a student of Fowler, and received the Heisenberg manuscript from him. He would later recall [58]:

> The first I heard of it was in September when Fowler sent to me a copy of the proofs of Heisenberg's paper and asked me what I thought about it. That was the first that I heard about it. I think Fowler found it interesting. He was a bit uncertain about it and wanted to know what my reaction to it would be. When I first read it I did not appreciate it. I thought there wasn't much in it and I put it aside for a week or so. Then I went back to it later, and suddenly it became clear to me that it was the real thing. And I worked on it intensively starting from September 1925.... I had been trying hard for two years to solve a certain problem without any success.

> Then I suddenly saw that Heisenberg's idea provided the key to the whole mystery.

Dirac recalled that he had seen something like the commutator relations in the guise of the Poisson brackets in classical physics. He was already familiar with matrix mathematics. He then developed essentially the same ideas that appear in the Born–Jordan paper. In essence, he confirmed the results of the latter paper, although he was unaware of it. He wrote up his paper and submitted it to an English journal [59]. He then sent a copy to Heisenberg, who pointed out to him that similar work was in the Born–Jordan paper as well as in a new manuscript that he was working on with Born and Jordan. Dirac's manuscript was received by the editors on November 7, while the new manuscript mentioned by Heisenberg reached their own editors on November 16. More to the point, Dirac's paper was published in December, while the Göttingen paper didn't make it into print until the following August. Dirac then went on to see if the new formulation would give the right properties of the hydrogen atom itself, which may have been the first good application of the new theory [60], with the paper appearing the following March. About this time, Pauli also used the new quantum mechanics to describe the hydrogen atom [61].

Dirac referred to the quantum variables as "q-numbers," as opposed to the classical quantities which were referred to as c numbers. That is, the latter are just ordinary numbers, whereas the quantum variables are operators. This terminology would come to be used by many workers in the field, and he then wrote a paper on the algebra of these q numbers [62], and then set out to develop the theory further [63]. This latter paper would introduce the ideas of identical particles, symmetric and anti-symmetric wave functions and Fermi–Dirac statistics. It too would appear in August 1926. Here, he would say,

> One can build up a theory without knowing anything about the dynamical variables except the algebraic laws that they are subject to, and can show that they may be represented by matrices whenever a set of uniformising [*sic*] variables for the dynamical system exists.

Hence, Dirac took the original idea of Heisenberg about the importance of experimental observables and replaced it with the importance of a proper mathematical theory. Here, reality existed from the basis upon which the theory was produced. Moreover, Dirac had made significant advances in the foundations of quantum mechanics, and particularly in the mathematical basis of the theory, while the Göttingen group was waiting for their paper to appear in print. In fact, Dirac was the first to introduce a wave function in the form of a determinant, which (later) would more commonly be called the Slater determinant [64]. The result of this is [65]

> it was recognized that the essence of the [exclusion] principle is the requirement that all state functions, with the inclusion of the spin functions, of a system of similar particles, provided they are fermions, must be antisymmetric with respect to particle exchange.

This led to Holton's comment on Dirac versus Heisenberg [48], and the fact that "a theory that has some mathematical beauty is more likely to be correct than an ugly one that gives a detailed fit to some experiments," which was cited above.

Margenau has raised the question as to why the exclusion principle formulated by Pauli and mathematically expressed via the antisymmetric wave functions by Dirac has not received more attention from the philosophers [66]. As he points out, the antisymmetric wave function means that it is not decomposable into a product of wave functions, one for each particle. This, in itself, "implies the presence of interactions, i.e. forces." The interaction in this case lies in the Coulomb forces between the electrons, but is expressed through the antisymmetry. In the simplest interpretation, the act of interchanging two identical particles by switching their positions means that the sign of the net wave function must change. This interchange is referred to as the exchange operation. However, this is a simple view, and, as Rosenfeld says [20], "The exclusion principle ... gives rise to an additional interaction ... called exchange interaction to remind us that it has its origin in the identity of the particles, and not at all because it could be attributed to any actual permutation of their locations."

The Drei–Männer Paper

As we have seen, Born, Heisenberg, and Jordan were working hard on a longer exposition to follow up on their first two papers. The paper was referred to by Heisenberg as the "dreimännerarbeit" and this name has kind of stuck to it. By November, they had more or less completed the paper, and it was accepted by the journal on November 16, 1925, but as we remarked it did not appear until the following August [67]. From the very name of the paper, it was clear that this was a sequel to the Born–Jordan paper, and extended the work to an arbitrary finite number of degrees of freedom. By this approach, they too were able to work with the many-particle picture. As Dirac had found, they too would find the proper antisymmetric wave functions to account for the interchange of the position of two particles. In addition, they worked through the entire formulation of perturbation theory so as to apply the results to more complicated problems. In the matrix formulation, the perturbation series was described via a scattering matrix, often referred to as the S matrix. The use of a preferred set of basis functions was augmented by a transformation theory which utilized unitary matrices to implement basis set transformations. This means that one can always find a system in which the Hamiltonian approach yields a diagonal matrix for the energies, and therefore the energy eigenvalues can be used to find a orthonormal basis set of eigenstates for these energies. Today, we know the diagonalization procedure as a normal part of the theory of linear algebra. This paper became the cornerstone of the most extended understanding of the new quantum mechanics, at least as it was espoused by the Göttingen crowd and their colleagues. Nevertheless [68], "the matrix form of quantum mechanics had been put into shape as a rigorous and powerful theory, ready to tackle whatever problems would be presented to it."

Several important points arise in this paper. For example [67], "any distinction between 'quantized' and 'unquantized' motion ceases to be at all meaningful, since the theory contains no mention of a quantization condition which selects only certain types of motion from among a large number of possible types: rather, in place of such a condition one has a basic quantum mechanical equation which is applicable to all possible types of motion." Here, one finds the matrix element involving the frequency difference that arises from the energy difference between two states. The authors presume that there is both a continuum of energies as well as a discrete set of energies, the latter of which occur within the atoms. Section 3.3 of the paper deals explicitly with the matrix approach to continuous spectra. The continuum could account for unbound electrons, for example. Then, "in the present theory, however, the fundamental principles of quantum theory and the principle of conservation of energy follow mathematically from the quantum-mechanical equations." Thus is BKS (the Born paper discussed in the previous chapter) delivered a death blow beyond what the experiments had achieved. But then, the authors continue to introduce phraseology which will haunt the field for decades (and lead to serious debates on the subject), and that is the *quantum jump* as the supposed particle moves from one state to the other. Here, they remark, "For a quantum jump in which there is a change in the angular momentum . . . the 'plane of vibration' of the generalized 'spherical wave' lies perpendicular to the z-axis. For every quantum jump [the angular momentum] changes by 0, or by $\pm\hbar$. The light emitted in the latter case is circularly polarized." They summarize their results with the comments:

> In our theory, kinematics and mechanics have again been brought into as close a relationship as that prevailing in classical theory, and that the new fundamental viewpoints, stemming as they do from the basic postulates of quantum theory for the mechanical concepts together with the concepts of space and time, find adequate expression in kinematics just as in mechanics and in the connection between kinematics and mechanics.

In spite of the three authors all being associated with Göttingen [69], "Bohr was inclined to see Heisenberg's breakthrough as a product of his time in Copenhagen and the influence of both Kramers and Bohr." The fact that Heisenberg had not cited Bohr in his first work was taken as a slight by the latter, yet Bohr held out great hope for the success of this new quantum mechanics. Yet [69], "to Einstein, it was more like mathematical trickery that divorced itself from Cartesian space-time coordinates."

A few days after the completion of this paper, Born left for the United States. He had been invited by Norbert Wiener to be a "Foreign Lecturer" at MIT [70]. The latter was quite well versed in matrix mathematics and they would pursue work to clean up some of the loose ends in the matrix formulation of quantum mechanics, particularly the manner in which operators are introduced into the theory [71]. For his part, Heisenberg was back in Copenhagen, completing his fellowship, and would produce his second shot in the new year.

The Peacemaker

Paul Ehrenfest was still determined to try to calm the water between Bohr and Einstein. The rise of the new quantum mechanics had only served to stir up the debate once again, but Ehrenfest would not give up the task. His dream was to have the two together in Leiden in a relaxed environment. There was to be a symposium to honor the 50th anniversary of Hendrik Lorentz's doctorate, and of course the two principal scientists were invited. He invited them to spend several days at his home, so that they could enter into (he hoped) meaningful discussions that would clear the air between them. In fact, he encouraged Bohr [72]:

> I urgently request that you bring no sort of writing work with you. It would really be a sin to spoil this opportunity—this rare opportunity to peer for once with Einstein into the furthest depths of physics that are accessible at present to anyone's gaze. And you would certainly, and at the same time uselessly, spoil it for yourself, if you were to bring such work with you.

Ehrenfest was promising that the two of them could concentrate purely on science, with evening chats and daily strolls. They would be free from interruption. Ehrenfest even went so far as to give them the privilege of smoking in their guest rooms, an act which was normally banned by Ehrenfest's wife.

The Uhlenbeck–Goudsmit paper had just appeared and everyone was talking about spin. On his way, Bohr had met Pauli at the train station in Hamburg, and Pauli had asked him what he thought. Bohr replied with his traditional "interesting," which in Copenhagen speak often meant "nonsense" [73]. When Bohr finally arrived in Leiden, Ehrenfest and Einstein asked the same question, so he was forced to provide some discussion of his objections. It was hoped that the two great scientists could be allowed sufficient time alone to discuss their differences, but things kept getting in the way. Of course, Uhlenbeck and Goudsmit were invited to discuss their new ideas on spin, which to any young scientist must have felt like a trip to the inquisition.

Nevertheless, Bohr and Einstein found enough time to discuss the state of physics and to figure what they agreed upon and on what they differed in their viewpoints [73], "even though they did not find the common ground that Ehrenfest had hoped for." But the rest and relaxation was good for both of them, and while they spent their evenings in Leiden, a new wave was about to break upon them all.

The Waves Come Ashore

By 1925, Erwin Schrödinger had been in Zürich for several years. He had previously turned down a position in Vienna to take the one at the University of Zürich. His wife Anny (for Annemarie) has said [74],

> Because in Vienna there was a terrific financial situation you see. He asked for a salary which they couldn't pay.... And then came Zürich and that, of course, was absolutely wonderful for us, leaving the inflation country and coming to Zürich. In Zürich there were Weyl, Debye, Scherrer, Edgar Meyer—many, many people. We were all very good friends together and of course Weyl was very much interested in my husband's work.

At the University of Zürich, Schrödinger was basically alone. However, it was worthwhile because there were several good physicists at the Technical University, a federally supported institution. And, in the coming effort, he would meet and discuss with Hermann Weyl quite frequently. When de Broglie's paper appeared with his wave and particle concept, it was agreed among the group that they would ask Schrödinger to give a joint colloquium on the topic. Schrödinger had become familiar with the de Broglie work through its mention in a paper by Einstein in the popular press, where he indicated that perhaps de Broglie's approach was quite promising, and Schrödinger generally attributed this as his beginning of interest in wave mechanics. Schrödinger had already begun to work on his own version of the wave nature of quantum mechanics, but was focused upon incorporating Einstein's relativity theory into the result. His results were not promising at the time, as he could not produce the proper quantization. So, he gave up for several months, returning to the issue near the end of 1925 [75]. During the winter vacation period, Schrödinger spent his time at a resort cabin with his L'amour du jour, in Arosa, Switzerland. As Jones puts it [76],

> Erwin and Anny had plenty of romantic drama in their lives, just not with each other; and Erwin couldn't help but be annoyed by the fact that Anny was not the least bit concerned about his various liaisons. It was a double standard, of course, but he seemed to expect Anny to be accepting of his falling in love with any attractive woman who walked by while at the same time wanting her to be at least somewhat jealous of his attentions to other women.

Gribbon put it slightly differently [77]:

> Although Schrödinger had many affairs with women, these were seldom, if ever, casual relationships. Judging from his diaries, love was more important to him than sex, although often sex naturally had its place in a loving relationship. He was often in love—or convinced himself that he was in love—and when he was in love, by and large life was good and his scientific creativity benefited.
>
> Anny Schrödinger and Herman Weyl . . . soon became an item, as did Weyl's wife, Hella, and the physicist Paul Scherrer.

Zürich was seemingly a very modern society, preceding the "free love" approach of the Vietnam era by almost half a century. For Schrödinger, life was very good indeed that winter. During this stay, he developed the meat of four important papers establishing a new quantum mechanics based on continuous wave theory, as put forward by de Broglie. Herman Weyl would categorize this massive achievement as a [77] "late erotic outburst in his life."

In the first paper, Schrödinger developed the time-independent form of his wave equation and showed how the atomic spectra of the single electron atom was a natural eigenvalue problem in mathematics [78]. His famous equation in one spatial dimension is

$$-i\hbar\frac{\partial\psi(x, t)}{\partial t} = \frac{\hbar^2}{2m}\frac{\partial^2\psi(x, t)}{\partial x^2} + V(x)\psi(x, t), \qquad (3.12)$$

where $\psi(x, t)$ is the wave function and is a complex quantity. Just a few weeks later, the second paper appeared, in which he continued the treatment showing how the energies of the oscillator used by Planck would occur [79]. In the third paper, he established that the matrix mechanics of the Göttingen crowd was fully equivalent to his wave equation approach [80]. While he had introduced the time-dependent wave equation in his first paper, he returned to it in the fourth paper and showed how perturbation theory could be applied to problems that were more difficult to solve [81]. Schrödinger's work crashed upon the physics community as a tidal wave smashing the shore. Here was an approach that produced all the correct results for the new quantum mechanics, but was eminently easier to understand than the complex mathematics of the matrix theory. The continuity of the waves provided a more understandable approach to the new theories, and essentially all physicists were accepting of waves as they had experience with them in acoustics and electromagnetics. This was something they could get their head around. There were, however, some subtleties that underlay the extension from classical physics. For Schrödinger, his theory held that it was the waves that were fundamental, and particles could only exist as localized regions of the waves (essentially what we would call a wave packet).

But not all physicists were so accepting. Heisenberg wrote to Pauli to say, "The more I ponder the physical part of Schrodinger's theory, the more disgusting it appears to me" [46]. Heisenberg would apparently never accept the wave approach to quantum mechanics. Part of this reasoning was, according to Rosenfeld [82],

"Schrödinger's wave mechanics was a direct onslaught on quantum mechanics, on quantum theory [of matrix mechanics] altogether." On his part, Schrödinger was not so accepting of matrix mechanics and particularly did not like the quantum jumps required for this theory. He was not alone, as the previously cited comment from Enrico Fermi clearly shows. This dispute would drag on continuously, with no apparent resolution. Along this line, Bohr invited Schrödinger to visit Copenhagen and to lecture on his new wave mechanics. Schrödinger went to Copenhagen at the end of September (1926). Heisenberg was in Copenhagen as well at this time, and there were long and probably contentious discussions going on. While the three men remained friends, their views on quantum mechanics were not changed, and Schrödinger is said to have remarked [77],

> If all this damned quantum jumping were really here to stay, I should be sorry I ever got involved with quantum theory.

But Bohr would resort to his standard approach of late night visits and long discussions which would ultimately lead to Schrödinger becoming ill. He would leave and return to Zürich to recover.

Quantization is an eigenvalue problem. Schrödinger used various forms of this phrase as the title to at least four papers in 1926, three of which were in the important group of 4. The eigenvalue problem was known in classical mechanics for quite a long time already, as we discussed above in connection with Sommerfeld's advancement of the Bohr atomic theory. Work on this problem dates from at least the middle of the eighteenth century, although the phrase "eigenvalue" and "eigenfunction" probably date from the work of Hilbert at the start of the twentieth century. In general if we take a second-order differential equation such as setting the right-hand side of Eq. 3.12 to 0 (or to any constant), then the nature of the solutions will be determined by the function $V(x)$ and the boundary values that are invoked. For example, if we take a simple stretched string which is constrained at its two ends, then the boundary values are that the amplitude of motion is zero at the ends. The potential $V(x)$ is determined by the stretching force placed upon the string (the tension in the string). Now, this string will have a fundamental mode of vibration in which the amplitude varies as

$$\sin\left(\frac{\pi x}{L}\right), \tag{3.13}$$

where L is the length between the two ends of the string. This is known as the fundamental mode and has a wavelength $\lambda = 2L$. Of course, any harmonic of this can also occur, in which the argument of the sine function is multiplied by an integer n. These harmonics are often called *overtones*. The amplitude of the fundamental and the various harmonics are determined by just where the string is "plucked." As we know, the set of the fundamental and these harmonics form a Fourier series which describes the motion of the string, and the various coefficients in this series give the amplitude of each mode. The frequency of the oscillation is determined by the mass of the string and the tension in the string, and this frequency is related to the wavelength and the velocity of sound v_s through the relationship

$$v_s = \frac{\omega}{k} = \frac{2\pi f}{2\pi/\lambda} = f\lambda. \tag{3.14}$$

Sound waves are compressional waves, in which the air is alternately compressed and expanded as one moves along the position of the wave. For these waves, the velocity of sound is not constant, whereas light waves have a constant velocity given by the speed of light. Thus, a tree falling in the forest will create a sound wave whether or not anyone is there to hear it (sorry, Mach).

There are a vast number of different potential functions $V(x)$ which arise not only in one dimension, but also in three dimensions. The solution of the equation thus depends upon the exact form of the potential function as well as the boundary values that are imposed. For a linear, second-order differential equation such as the right-hand side of Eq. 3.12, two boundary values can be imposed. As mentioned earlier, the general form of this equation is often called the Sturm–Liouville equation after two French mathematicians, Jacques Sturm and Joseph Liouville, who did their work in the mid-nineteenth century. Various forms that arise from the coordinate system and the potential function have been found, and give rise to a wide library of special functions, usually named after the discoverer of that particular function. Discussions of the special functions and the nature of the solutions is reviewed in a great many books, perhaps the best being Morse and Feshbach [4] and the volumes arising from the Bateman project [5].

The unique importance of the solutions to the eigenvalue problem is that they form the basis for the mathematics of linear algebra. In general, the solution can be represented as a vector field. This may be confusing, but quite generally, a field is a function that has a specific value at each and every point in space. A vector field also has a direction at each and every point in space. To give this some substance, we normally think about a three-dimensional coordinate system. The directions of each of these three coordinates is defined by a unit vector, which has unit amplitude and points along one of the three axes. Then, a vector is defined as so many units along direction 1, so many units along direction 2, and so many units along direction 3. This idea can be extended to many more directions. For example, in the plucked string discussed above, the eigenfunction (Eq. 3.13) corresponds to the unit vector along the fundamental direction. Each of the overtone functions, which have Eq. 3.13 with an integer n in the argument, corresponds to other directions—the overtone directions. Then, the Fourier series representing the amplitude of the string at each position is a vector field, in which the coefficient of each term in the series is the amplitude of that coordinate. The set of sine functions form what is called a linear basis set. The set of coefficients give a matrix defined upon this basis set. Thus, the continuous wave for the string motion can be expanded as a matrix on this basis set, which is just the Fourier series, but the matrix connects to the Göttingen form of quantum mechanics. Hence, the simple three-dimensional space can be extended to multiple (and even infinite) dimensions through these basis sets. The normal Fourier series is such an infinite dimensional space. The multiple (or infinite) dimensional space described by a set of basis functions has come to be known as the Hilbert space [83]. It is important to note that this space, and these ideas existed long before quantum mechanics, and they are not unique to the latter. In fact, Hilbert himself, along with two of his young researchers (including John von Neumann), would later publish a mathematical treatment of quantum mechanics which formalized the use of linear vector spaces [84].

Another important aspect of the Schrödinger wave function is that now the group velocity of the wave could be related to a particle velocity, were the particle to exist. If the particle were a localized wave packet, then it would be natural that its velocity would be that of the group of waves. This would make a close connection between the particle of classical physics and the wave packet of quantum mechanics. Thus, it can be seen how many have said that Schrödinger long argued a connection between classical mechanics and quantum mechanics. As Rosenfeld remarked [85],

> Schrödinger made no secret of his intention to substitute simple classical pictures for the strange conceptions of quantum mechanics, for whose abstract character he expressed deep "aversion"; he was conscious that this last sentiment was shared by all the older generation of physicists who had not accepted the necessity of giving up their habitual ways of thinking when dealing with phenomenon on the atomic scale.

This is an important philosophical viewpoint, as it makes quantum mechanics less mysterious and confusing. The mathematics was essentially the same, with the principal new concept being that the particle was in reality a wave, as initially put forward by de Broglie. Here, Schrödinger and de Broglie would agree that the concept of a particle is better described in terms of a localized wave, which appears to a particle only when not viewed microscopically. The connection with classical physics was explored further by Erwin Madelung [86], who cast the Schrödinger equation into a hydrodynamic form. Here, he would identify the continuity equation for probability current. He assumed that the wave function could be written in the form (he used three dimensions, but we illustrate it here with only one spatial dimension, and we use a more modern form)

$$\psi(x, t) = f(x, t) \exp\{i[\omega t - S(x, t)/\hbar]\}, \qquad (3.15)$$

so that S corresponds to the classical action (energy times time). By separating the real and imaginary parts of Schrödinger's equation with this substitution, he was able to recover some of the hydrodynamic properties. In addition, he was able to show that every eigenfunction solution corresponded to a stationary flow pattern. As a result, he felt that there was a rationale to say that the quantum theory could be treated on the basis of hydrodynamics. That is, the wave flow was just not that different from the concept of fluid wave flow.

For his part, Born liked the mathematics of Schrödinger, as it was much in the form he viewed as being a very basic form. He remarked [87],

> Of the different forms of the theory only Schrödinger's has proved suitable for this [scattering] process, and exactly for this reason I might regard it as the deepest formulation of the quantum laws . . . the Schrödinger form allows one to describe not only stationary states but also quantum jumps.

This comment was a barb to Heisenberg. As he later remarked [42],

> I did not like so much that Born went over to the Schrödinger theory. Either that was just using a new mathematical tool instead of an old one for the same thing, in which case all right, why not do it from the old mathematical tool. Or it really meant that Born doubted that our scheme was right, and thought rather Schrodinger's scheme was right. Well, all right. That was, of course, an entirely new possibility and I realized that this was a possibility, but at that time I was not willing to accept this possibility.

Born went further in this paper, and here he formulated his probability conjecture that the magnitude squared of the wave function gave the probability of finding the "particle" at a point in space. Schrödinger had already shown that the magnitude squared of the wave function could be related to charge density and had developed the probability current [81]. Born went further with his probability interpretation. He recognized that, because the wave function was a complex quantity, the superposition of two wave functions would lead to an interference effect, by which the quantum probability would be significantly different than the classical probability. He pointed out that the wave function did not give the exact position of any particle, only that it contributed to the probability density of where the particle may be found. Thus, a measurement would give the actual value and do away with the probability, which meant doing away with the wave function. This idea leads to a so-called collapse of the wave function during the measurement, a point which would be amplified and argued over many times during the coming years. Born would eventually be awarded a Nobel prize for this new viewpoint. Somewhat later, Louis de Broglie extended this probability idea and considered how the "particle" could be related to a singularity in the wave field. He first tried this for photons, and then extended it to normal particles, saying [88], "continuous solutions of the wave equation provide merely statistical information; an exact microscopic description undoubtedly necessitates the usage of singularity solutions reflecting the discrete structure of matter and radiation." Thus, in de Broglie's view, there are two wave solutions, one is a continuous solution related to the probability and a second singular solution which would relate to a particle's position [89]. It is probably worth mentioning that Einstein rejected the statistical interpretation, and sent a short note to Born about it. Born's comments about the note conclude [52]:

> The note showed that Einstein rejected the statistical interpretation of quantum mechanics not just because of his "inner voice." He had tried a different, non-statistical interpretation of Schrödinger's wave mechanics and was submitting a paper about it to the Academy. I can't remember it now; like so many similar attempts by other authors, it has disappeared without a trace.

But Born would also push forward a new concept of determinism, removing it from its classical interpretation. In a talk later that year in Britain, he stated [90], "We free forces of their classical duty of determining directly the motion of particles and allow them instead to determine the probability of states. Whereas before it was our purpose to make these two definitions of force equivalent, this problem has now no longer, strictly speaking, any sense." The question of determinism would fester and grow.

Not only did we now have wave mechanics as a counterpoint to Heisenberg's particle matrix mechanics, the new wave mechanics led to new arguments arising from Bohr's need to remove causality. As Margenau would put it [66],

> the new theory functions precisely as did Newton's laws of motion in ordinary mechanics. These are essentially artefacts for predicting future states when present or past states are given.... Quantum mechanics provides equations (in particular Schrödinger's equation; in view of the customary designation given to Newton's discoveries, it should also be called a *law*) which permits the same convention, but relative to the newly defined states.

Thus, causality and determinism can be properly envisioned to apply within the new wave mechanics of Schrödinger. This would be a contentious view, and, as we shall see, would be deemed wrong by Göttingen and Copenhagen.

In 1927, Max Planck would retire from his position at the university in Berlin, and Erwin Schrödinger was invited to come to Berlin to fill his position. It is interesting that, with all the young lions in Germany, the university decided to bring Schrödinger from Zürich, perhaps sharing Born's view about the new wave mechanics being superior to the difficult matrix mechanics of Göttingen. Anny viewed the move favorably [74]:

> When we came to Berlin we stayed in Planck's house for the first time ... we thought, "Well, we will stay in Berlin for a good while." We couldn't have thought that we'd have to leave Berlin.
> "Berlin was the most wonderful and absolutely unique atmosphere for all the scientists. They knew it all and they appreciated it all."

Moreover, Einstein was in Berlin!

Heisenberg's Uncertainty

In late 1926, Heisenberg was again in Copenhagen, and had been there during Schrödinger's visit and arguments with Bohr. He recognized that the root of the problem perhaps lay in the fact that there was not a definitive interpretation of the quantum mechanical formulations that had been given. He had his own views about where the crisis was to be found. According to Pais [91],

> Bohr was trying to allow for the simultaneous existence of both particle and wave concepts, holding that, though the two were mutually exclusive, both together were needed for a complete description of atomic processes. Heisenberg thought otherwise.

As we remarked earlier, Heisenberg was having trouble with the concept of position or of angular momentum without these quantities having been measured. As McEvoy put it [92], Heisenberg

> criticized the formal quantization rules ... because they used concepts such as the position ... and period of revolution [angular momentum] to derive the observable energy levels of the atom. These ... were "apparently unobservable in nature".... Heisenberg objected to this on the grounds that the rules used ... are not even internally consistent or applicable to a clearly defined range of quantum mechanical problems.

Heisenberg himself described this period, and his development of the uncertainty principle [93]:

> Bohr and I tried from different angles and therefore it was difficult to agree. Whenever Bohr could give an example in which I couldn't find the answer, then it was clear that we had not understood what the actual situation was ... we both were in a kind of despair. In some way we couldn't agree and so we were a bit angry about it. So Bohr went away to Norway to ski. Earlier he had thought of the possibility of taking me with him but then he didn't like it. He wanted to be alone, to think alone, and I think he was quite right. So I was alone in Copenhagen and then within a few days I thought that this thing with the Uncertainty Relations would be the right answer. I tried to say what space meant and what velocity meant, and so on.... Then I found very soon that these are situations in which there was this Uncertainty Relation between p and q.

Heisenberg discussed his ideas with Pauli, as was his normal approach. Pauli was enthusiastic, which was not his norm. Heisenberg then wrote up the ideas in a manuscript. There is some debate as to whether he submitted the manuscript prior to showing it to Bohr, but it is likely that he first showed it to Bohr upon the latter's return from his skiing trip. But Bohr had some concerns and wanted changes, which Heisenberg resisted. In any case, the paper on uncertainty was received by the journal on March 23 and published in the March 1927 issue [94]. This was a very rapid turnaround of the paper.

Now, uncertainty was certainly not a new idea. It had existed in classical mechanics since the ancients. But, in classical mechanics, it was a problem of measurement. Anyone who has sat patiently (or impatiently) at a railroad crossing waiting for the train to pass has perhaps considered the problem of how to measure the length of a given car or the velocity with which it is moving. If you are alone, how do you fix the positions of the two ends of a selected car at a common time? Perhaps you draft a colleague to watch one end and you chose the other. How do you arrange so that you both mark the position of the two ends at a given time? Thus, your measurement of the position is, at best, quite inaccurate. As a result, any guess at the velocity of the train is also subject to this error since you need to know how the ends of the car move with time. So, this is inaccurate. With modern science, you may say, "But wait. Can't I use modern radar to determine the position of the particle by the reflected wave and the velocity by the Doppler shift of the reflected wave?" Of course, within limits you can do this on the moving train car, but the microwave signal has a certain beam width, which introduces an error in the position of the object producing the reflection. And the measurement of the reflected frequency, due to the Doppler shift, has an uncertainty due to the experimental apparatus. Thus, I likely have reduced my measurement uncertainty, but I have not gotten rid of it. While the Doppler effect predates relativity, the latter changes the lengths and frequencies of the moving object and further degrades my ability to determine position and velocity (momentum). And then of course, I have to assume that no one is "messing with" my Radar signal. In the military, there are approaches known as countermeasures by which the intent is to make me think that the object generating the return signal is somewhere other than where it really is. So, the measurement accuracy may be degraded by active interference. Of course, there are counter-countermeasures to prevent this from interference from occurring, but this is a vicious circle producing an infinity of levels, much like sitting in an older barber shop with the walls covered by mirrors so that one can observe an infinity of reflections. The important point of all of this wandering away from the question is to realize that, in classical physics, the uncertainty is a result of our inability to make an accurate measurement. But then in statistical physics, such as in Brownian motion, the best I can do is measure an average for an ensemble of objects, since the actual position and velocity of any small particle is beyond my mathematics to define, which is why I am using statistical physics in the first place.

So, we have to expand the concept of uncertainty to either measurement error or to the fact that the actual position and velocity simply are variables which are hidden from me by the physical description with which I am working. This idea of "hidden variables" will reoccur often in the development of quantum mechanics. Many of the principle devotees would devote considerable effort to show that such hidden variables were not possible, but most of these efforts would prove to be no more than tautological arguments.

Heisenberg went further. He said that the position and momentum of the particle were, in fact, unknowable in quantum mechanics. And, because of this unknowable nature, it would be impossible to measure them, no matter the accuracy of the measuring apparatus. One way to think about this is via de Broglie's waves. If we have a wave of a fixed wavelength, which for a photon gives a fixed frequency and momentum, then it has equal amplitude everywhere. While I know its momentum, I have no idea just where the photon is. If I go further and try to describe a particle by a wave packet, which is more localized in space, this packet has to be constructed of an ensemble of waves, each of which has a slightly different frequency. Hence, while I get a better estimate of position, I have given up knowing exactly what the frequency and momentum are. A common way to describe a wave packet is to assume that the various momentum components are described by a Gaussian distribution function. A Gaussian has a well-defined standard deviation, which is a root-mean-square (RMS) estimate. It is found by finding the average value $\langle p \rangle$, given by the peak of the Gaussian, and then determining the second variation $\langle p^2 \rangle$ and writing the standard deviation as $\sigma_p = \sqrt{\langle p^2 \rangle - \langle p \rangle^2}$. This Gaussian now produces an equivalent Gaussian for the position of the particle, with its own standard deviation σ_x. Heisenberg goes on to say that these two standard deviations must satisfy the relation

$$\sigma_x \sigma_p \geq \frac{\hbar}{2}. \tag{3.16}$$

In classical physics, x and p are considered to be conjugate variables, as they are the natural variables for motion in one spatial dimension, and the Hamiltonian for the energy would be written in terms of these two variables. In quantum mechanics, whether for the wave mechanics or the matrix mechanics, these two variables become operators which do not commute. Thus, the order becomes important and using xp gives different results than using px. The two variables do not commute, and matrices corresponding to them do not commute, which was the basis for Heisenberg's formulation of matrix mechanics. What it means in terms of the Hilbert eigenfunction type of argument is that since they do not commute, they will not have the same set of basis functions, so that their eigenvalues cannot be determined simultaneously. As result, it becomes impossible to estimate these eigenvalues for both operators at the same time, and they are "unknowable" in Heisenberg's terminology. As Stephen Brush has said [95],

> Quantum physics had to take account, on a very fundamental level, of the fact that the act of observation disturbed what one is observing, so that no measurement can be completely accurate. The physicists put the blame for this failure on nature, and called it the uncertainty principle.

Heisenberg would consider that any use of the position and momentum as descriptors of a particle would be meaningless if they violated his uncertainty principle. But, philosophically, he would go further, saying [94],

> When one wants to be clear about what is to be understood by the words "position of the object," for example of the electron . . . then

one must specify definite experiments with whose help one plans to measure the "position of the electron"; otherwise this word has no meaning.

Now, we see the Bohr orthodoxy appearing here, as it did in Heisenberg's first paper on matrix mechanics. To understand quantum phenomena, one needs both the concept and the explicit experimental arrangement by which the concept is to be observed. Because of this, he can rule out a specific designation of an orbit for an electron in the atom [94]:

> We now turn to the "path of the electron." By path we understand series of points in space . . . which the electron takes as "positions" one after another . . . the often used expression, the "1s orbit of the electron in the hydrogen atom," from our point of view has no sense. In order to measure this 1s "path" we have to illuminate the atom with light whose wavelength is considerably shorter than 10^{-8} cm. However, a single photon of such light is enough to eject the electron completely from its "path". . . . Therefore here the word "path" has no definable meaning.

We may conclude, then, that, within this view, there is no reality to the 1s orbit without the observation via experiment. Hence, there is no meaning in discussions of this orbit within the atom. In the paper's conclusions, Heisenberg also feels compelled to strike out against causality [94]:

> But what is wrong in the sharp formulation of the law of causality, "when we know the present precisely, we can predict the future," is not the conclusion but the assumption. Even in principle we cannot know the present in all detail. For that reason everything observed is a selection from a plenitude of possibilities and a limitation on what is possible in the future. As the statistical character of quantum theory is so closely linked to the inexactness of all perceptions, one might be led to the presumption that behind the perceived statistical world there still hides a "real" world in which causality holds. But such speculations seem to us, to say it explicitly, fruitless and senseless. . . . Because all experiments

are subject to the laws of quantum mechanics . . . it follows that quantum mechanics establishes the final failure of causality.

But this hides his own assumptions, and the argument is something of a tautology. He has assumed that there is no underlying reality other than what can be determined by the experiments, a quite Machian assumption. Because there is no underlying reality, there of course is no hidden "real" world, and he can draw these conclusions. In some sense, here, he is fighting against Schrödinger, a fight he brings up many times in the paper, because the wave equation formulation of Schrödinger gives life to an underlying reality, much as Margenau expressed it at the end of the last section. If the 1s orbit exists as Schrödinger would tell us, even though it may not be measurable, then there remains a sense of underlying reality. This argument about the interpretation of quantum mechanics would not go away.

Complementarity

The year 1927 had started in an auspicious manner with Heisenberg's uncertainty paper appearing in March. Erwin Schrödinger was now ensconced in Berlin as the successor to Planck. Heisenberg had turned down an offer to come to Leipzig as professor in order to become Bohr's second in command in Copenhagen, which allowed him to formulate his important paper on uncertainty, but at the moment he had no prospects for a professorship, although Leipzig would approach him again later in the year. And Niels Bohr had a problem. The central axis for the development of quantum theory had shifted from Copenhagen to Göttingen. In fact, all of the developments for the new quantum theory had largely come from the Germanic scientists. His own atomic theory was now passé. Even when proposed, there were mathematical and physical problems with it, and Sommerfeld had corrected the mathematics. His BKS paper (with Kramers and Slater), with its new ideas, had been shown to be wrong almost as soon as it was published. Now, he had done little in the intervening years to re-establish Copenhagen as the shining light of the new theory. It was true that Heisenberg had chosen to come to Copenhagen as his new assistant, but this had led to continuing arguments over the interpretations to be applied to the new quantum mechanics. In fact, he had found an error in one of Heisenberg's examples in the uncertainty paper, but Heisenberg had refused to recall the paper to correct it. He had added a short note on Bohr's views, but the main paper remained unchanged. In short, Niels Bohr was becoming irrelevant. If he wanted to remain the high priest of quantum mechanics, he needed to do something quickly.

It was already known that there would be a major conference, with limited attendance, in Brussels later in the year. The invitations had gone out already in January, and it was clear that it would be a contest over the soul of quantum mechanics. But he could not wait until then, because one could not count on a given outcome with so many leading scientists present. He had been invited to a conference celebrating the centenary of the death of Volta, to be held in Como, Italy, that September. He had actually been invited already the previous year, but he ignored the invitation. But he was expected to take part on behalf of the Danish academy. In his added note, Heisenberg remarked [94], "Bohr ... sharing with me at an early stage the results of these more recent investigations of his— to appear soon in a paper on the conceptual structure of quantum theory." So, it was already clear in March that Bohr had decided to make the Como lecture the highlight of his views on quantum mechanics. In fact, Bohr had already noted in letters to Fowler and to Einstein that he intended to talk about the general principles of quantum theory [96]. Certainly, many of his colleagues would be present as well. Bohr's talk eventually became known as where he stated his principle of complementarity. But Bohr's well-known problem with writing meant that the manuscript was not ready by the time of the conference. Pauli would stay after the conference with Bohr and help to get the final version ready, but it would not be published until the next year [97], and not at all in the conference proceedings.

Bohr intended to take the new quantum mechanics as a basis for an entirely new approach which would provide the philosophical underpinnings of this new theory. As Holton has said [98],

Bohr's proposal of the complementarity principle was nothing less than an attempt to make it the cornerstone of a new epistemology ... the situation in quantum physics is only one reflection on an all-pervasive principle ... it was the universal significance of the role of complementarity which Bohr came to emphasize.

Up until now, it was said that [99]

Bohr himself never stated a full philosophical position on any of the topics which were central to him ... [and] never worked out the details necessary to fit these philosophical components into a sufficiently explicit epistemological system to support his ideas.

Now, however [99],

Bohr felt more deeply than his contemporaries the need to understand the implications of the new theories on quantum ideas. He experienced the failure of attempts to visualize the processes underlying these theories as a severe dislocation and came to realize that this situation required a philosophical analysis if we are to understand the proper scope and application of these theories.... Bohr's contributions, to both physics and epistemology, were central to the development and interpretation of quantum mechanics as it emerged.

So, complementarity was to be the central tenet of his new philosophy, one which he hoped would be applicable to all fields of science—his new theology for science.

This manifesto would clarify just where he stood on quantum theory, whether one was talking about the old version or the new version, and how many classical ideas would no longer be applicable in this new philosophy. He laid the gauntlet down early with the statements [97]

the quantum theory ... essence may be expressed in the so-called quantum postulate, which attributes to any atomic process an essential discontinuity ... completely foreign to the classical theories and symbolized by Planck's quantum of action.

> This postulate implies a renunciation as regards the causal space-time co-ordination of atomic processes.... Accordingly, an independent reality in the ordinary physical sense can neither be ascribed to the phenomena nor to the agencies of observation.
>
> ... if in order to make observation possible we permit certain interactions with suitable agencies of measurement, not belonging to the system, an unambiguous definition of the state of the system is naturally no longer possible, and there can be no question of causality in the ordinary sense of the word.

Here, we have it in a nutshell. Reality exists only in the measurement, which itself leads to collapse of the wave function, and loss of the "definition of the state" of the quantum system in which we are interested. Causality has been completely done away with.

The importance of the recent Göttingen papers, particularly those of Heisenberg, are then discussed and their importance emphasized. But he does not discard wave mechanics, and strives to bring de Broglie and Schrödinger into the fold. Yet, even here, he tilts toward supporting Heisenberg [97]:

> An adequate representation of the stationary states ... entails ... in the interpretation of observations a fundamental renunciation regarding the space-time description is unavoidable. In fact, the consistent application of the concept of stationary states excludes ... any specification regarding the behavior of the separate particles in the atom.

Bohr begins to give a better definition of complementarity with the observation that [97]

> the fulfillment of the claim of causality for the individual light processes ... entails a renunciation as regards space-time description. Of course, there can be no question of a quite independent application of the ideas of space and time and of causality. The two views of nature of light are rather to be considered as different attempts at an interpretation of experimental evidence in which the limitation of the classical concepts is expressed in complementary ways.

... it follows from the above considerations that the measurement of the positional co-ordinates of a particle is accompanied not only by a finite change in the dynamical variables, but also the fixation of its position means a complete rupture in the causal description of its dynamical behavior, while the determination of its momentum always implies a gap the knowledge of its spatial propagation ... this situation brings out most strikingly the complementary character of the description....

In fact, wave mechanics, just as the matrix theory, ... represents a symbolic transcription of the problem of motion of classical mechanics adapted to the requirements of quantum theory and only to be interpreted by an explicit use of the quantum postulate. Indeed, the two formulations of the interaction problem might be said to be complementary in the same sense as the wave and particle idea in the description of the free individuals.

Here, then, is the essence of the complementarity principle. It would be debated and reinterpreted for decades. Heisenberg rebelled at the inclusion of a meaningful place for wave mechanics. McEvoy remarks [100],

It [complementarity] sparked significant debate in the years that followed and solidified the boundaries between those who accepted Bohr's view of the consequences of quantum theory and those ... seeking a more "realistic" microscopic theory or ... interpretation.... Bohr was accused of being obscurantist and pessimistic by setting boundaries to where science could go.

Out of these discussions came a rough consensus—accepted by a number of workers but bitterly opposed by some—which was soon called the Copenhagen Interpretation [CI] of quantum mechanics ... it was clearly Bohr who was most concerned with ... the philosophical issues underlying ... quantum mechanics.

While the complementarity ideas formed the basis for what is often called the Copenhagen interpretation, as mentioned above, it is clear that not even all of Bohr's colleagues agreed with him. Certainly Heisenberg disliked any mention of wave mechanics, and Born had differences with Bohr on philosophy and the role of mathematics. Although Born had originally endorsed wave mechanics, he too would gradually turn away from it in favor of the matrix formulation.

The talk itself was a disaster. At the best of times, Bohr was a confusing speaker, as we have remarked in earlier chapters. In general, most of the scientists present had a great deal of difficulty in following Bohr's often confusing arguments (the written manuscript is not significantly better in this regard). Many considered both the talk and the paper totally incomprehensible. And, Heisenberg was probably not happy with Bohr indicating that the Göttingen research was an outgrowth of Bohr's own research [101]. Like the BKS paper, the Como talk landed with a thud. It was not reported much in the scientific press. As Jammer remarks [102], "Physicists engaged in the application of the new formalism ... were too busy to direct their attention to questions of interpretation, and philosophers were generally still lacking technical knowledge to participate in the debate." Moreover, Bohr himself did not define just what he meant by "complementarity." It remained for others to define the phrase in a way that they supposed Bohr would have meant. Bohr, for whom the use of an exact phrase was crucial in his writing, thus failed himself at a crucial point.

In general, though, the Como conference was to be a precursor to Solvay to be held the next month in Brussels. In Como, Heisenberg, Schrödinger, Pauli, and de Broglie talked. The discussion was pleasant enough, but the ideas were planted for dissension to arise when next they met. Yet, Bohr was here laying a claim to the intellectual understanding and the philosophical interpretations to be assigned to the new quantum theory. And, it was clear that this understanding and interpretation were to be very different from the reality view of the world that Einstein, Schrödinger, de Broglie, and a significant number of the scientists of the day, all held dear to their hearts. If Bohr were to have his way, they were entering a brave new world. But this world would not be as all pervasive as Bohr had hoped. He had hoped to give a new epistemological theory which could be extended beyond quantum mechanics. In the end, he was ignored by most philosophers outside those dealing mainly with physics.

Nevertheless, Bohr was defining complementarity as the purveyor of the truth of quantum mechanics. And this truth was to be carried by the Copenhagen and Göttingen constabularies. Those views held by Einstein, de Broglie, Schrödinger, and others who professed to believe in reality prior to any actual measurement, were to be sent to the dust bin. It is absolutely clear that Bohr was trying to invoke this new philosophy by dictatorial decree, just as he tried to impose proof by intimidation upon visitors and workers in Copenhagen. Perhaps, it has been put best by Norman Levitt in his review of James Cushing's book [103], where he offers the view [104]:

> In trying to account for the Copenhagen interpretation, many suggestions have been made. However, some note might have been taken of the role of sheer ambition. It may well be that the Copenhagen School's condemnation-in-advance of any attempt to restore realism and determinism to microphysics is best explained by an insight of Francis Bacon (The New Organon, Book I, Aphorism 88): They did it "all for the miserable vainglory of having it believed that whatever has not yet been discovered and comprehended can never be discovered and comprehended hereafter."

References

1. V. Weisskopf, in *The Nature of the Physical Universe: 1976 Nobel Conference*, ed. by D. Huff and O. Prewett (Wiley-Interscience, New York, 1979), p. 3.
2. Letter from P. Ehrenfest to H. A. Lorentz of 25 August 1913, cited in M. J. Klein, in *The Lesson of Quantum Theory*, ed. by J. de Boer, E. Dal, and O. Ulfbeck (North-Holland, Amsterdam, 1986), pp. 325–342.
3. I. S. Newton, *Philosophiae Naturalis Principia Mathematica* (S. Pepys Press, London, 1686).
4. P. M. Morse and H. Feshbach, *Methods of Theoretical Physics* (McGraw-Hill, New York, 1953), Chapter 6.
5. See, e.g., A. Erdélyi, ed., *Higher Transcendental Functions* (Krieger, Malabar, FL, 1953, 1981) these volumes are based upon notes left by H. Bateman, and found after his death; they were earlier called the Bateman manuscripts.

6. N. Bohr, *Philos. Mag.*, Ser. 6, **26**, 1 (1913).

7. H. Nagaoka, *Philos. Mag.*, Ser. 6, **7**, 445 (1904).

8. M. Jammer, *The Conceptual Development of Quantum Mechanics* (McGraw-Hill, New York, 1966), p. 91.

9. A. Sommerfeld, *Ann. Phys.* **51**, 1 (1916).

10. A. A. Michelson, *Philos. Mag.*, Ser. 5, **34**, 280 (1892).

11. P. Zeeman, *Philos. Mag.*, Ser. 5, **43**, 226 (1897); **44**, 55 (1897).

12. P. Zeeman, *Nature* **55**, 347 (1897).

13. J. Larmor, *Philos. Mag.*, Ser. 5, **44**, 503 (1897).

14. T. Preston, *Trans. R. Dublin Soc.*, Ser. 2, **6**, 385 (1986).

15. Letter from H. Ehrenfest to A. Sommerfeld of 10 May 1918, cited by M. J. Klein (see Ref. 2).

16. O. Stern and W. Gerlach, *Z. Phys.* **8**, 110 (1922); **9**, 349 (1922).

17. M. Jammer, *op. cit.*, p. 133.

18. T. E. Phipps and J. B. Taylor, *Science* **44**, 480 (1926).

19. T. E. Phipps and J. B. Taylor, *Phys. Rev.* **29**, 309 (1927).

20. L. Rosenfeld, in *Physics in the Sixties*, ed. by S. K. Runcorn (Oliver and Boyd, 1963), pp. 1–22, reprinted in *Selected Papers of Leon Rosenfeld*, ed. by R. S. Cohen and J. J. Stachl (Riedel, Dordrecht, 1979).

21. W. Heisenberg, The American Institute of Physics history program: https://www.aip.org/history-programs/niels-bohr-library/oral-histories/4661-4

22. S. Jones, *The Quantum Ten* (Thomas Allen, Toronto, 2008), p. 130.

23. E. C. Stoner, *Philos. Mag.*, Ser. 6, **48**, 719 (1924).

24. A. Landé, *Z. Phys.* **16**, 391 (1922).

25. W. Pauli, *Z. Phys.* **31**, 765 (1925).

26. J. A. R. Newlands, *Chem. News* **10**, 94 (1864); **12**, 83 (1865).

27. D. Mendeleev, *Z. Chem.* **12**, 405 (1869).

28. H. G. J. Moseley, *Philos. Mag.*, Ser. 6, **26**, 1024 (1913).

29. N. Bohr, *Nature* **107**, 104 (1921).

30. N. Bohr, *Philos. Mag.*, Ser. 6, **26**, 472 (1913).

31. W. Heisenberg, The American Institute of Physics history program: https://www.aip.org/history-programs/niels-bohr-library/oral-histories/4661-6

32. M. Jammer, *op. cit.*, p. 146.

33. A. L. Parson, *Smithsonian Miscellaneous Collections* **65**, 11 (1916).

34. N. Bohr, *Ann. Phys.* **71**, 228 (1923).

35. G. Uhlenbeck and S. Goudsmit, *Naturwiss.* **13**, 953 (1925).

36. A. Whitaker, *Einstein, Bohr and the Quantum Dilemma* (Cambridge Univ. Press, Cambridge, 1996), p. 131.

37. L. H. Thomas, *Nature* **117**, 514 (1926).

38. A. H. Compton, *Phys. Rev.*, Ser. 2, **14**, 20 (1919); **14**, 247 (1919).

39. W. Heisenberg, The American Institute of Physics history program: https://www.aip.org/history-programs/niels-bohr-library/oral-histories/4661-9

40. W. Heisenberg, The American Institute of Physics history program: https://www.aip.org/history-programs/niels-bohr-library/oral-histories/4661

41. S. Jones, *op. cit.*, pp. 160–163.

42. W. Heisenberg, The American Institute of Physics history program: https://www.aip.org/history-programs/niels-bohr-library/oral-histories/4661-7

43. W. Heisenberg, *Z. Phys.* **33**, 879 (1925), trans. in B. L. van der Waerden, *Sources of Quantum Mechanics* (North-Holland, Amsterdam, 1967), pp. 261–276.

44. M. Jammer, *op. cit.*, p. 198ff.

45. P. McEvoy, *Niels Bohr: Reflections on Subject and Object* (MicroAnalytix, San Francisco, 2000), p. 59.

46. G. Holton, *Thematic Origins of Scientific Thought: Kepler to Einstein*, 2nd (revised) edition (Harvard Univ. Press, Cambridge, 1988), pp. 15–16.

47. M. A. Tuve, in *The Search for Understanding*, ed. by C. P. Haskins (Carnegie Institute, Washington, D.C., 1967), p. 46, cited in Ref. 43.

48. G. Holton, *Thematic Origins of Scientific Thought: Kepler to Einstein* (Harvard Univ. Press, Cambridge, 1973), p. 26.

49. M. Born, The American Institute of Physics history program: https://www.aip.org/history-programs/niels-bohr-library/oral-histories/4522-1

50. B. L. van der Waerden, *Sources of Quantum Mechanics* (North-Holland, Amsterdam, 1967), p. 37.

51. M. Born and P. Jordan, *Z. Phys.* **33**, 479 (1925).

52. M. Born, *The Born-Einstein Letters*, trans. I. Born (Macmillan, London, 1971).

53. H. Goldstein, *Classical Mechanics* (Addison-Wesley, Reading, MA, 1950), Chapter 8.

54. M. Born and P. Jordan, *Z. Phys.* **34**, 858 (1925), trans. in B. L. van der Waerden, *Sources of Quantum Mechanics* (North-Holland, Amsterdam, 1967), pp. 277–306.

55. W. Heisenberg, in a letter to W. Pauli, October 12, 1925, cited in S. Jones, Ref. 39.

56. W. Heisenberg, The American Institute of Physics history program: https://www.aip.org/history-programs/niels-bohr-library/oral-histories/4661-5

57. N. Bohr, The American Institute of Physics history program: https://www.aip.org/history-programs/niels-bohr-library/oral-histories/4517-2

58. P. A. M. Dirac, The American Institute of Physics history program: https://www.aip.org/history-programs/niels-bohr-library/oral-histories/ 4575-1

59. P. A. M. Dirac, *Proc. R. Soc. London, A* **109**, 642 (1925).

60. P. A. M. Dirac, *Proc. R. Soc. London, A* **110**, 561 (1926).

61. W. Pauli, *Z. Phys.* **36**, 336 (1926).

62. P. A. M. Dirac, *Proc. Cambridge Philos. Soc.* **23**, 412 (1926).

63. P. A. M. Dirac, *Proc. R. Soc. London, A* **112**, 661 (1926).

64. J. C. Slater, *Phys. Rev.* **34**, 1293 (1929).

65. M. Jammer, *op. cit.*, p. 144.

66. H. Margenau, *Philos. Sci.* **11**, 187 (1944).

67. M. Born, W. Heisenberg, and P. Jordan, *Z. Phys.* **35**, 557 (1926), trans. B. L. van der Waerden, *Sources of Quantum Mechanics* (North-Holland, Amsterdam, 1967), pp. 321–385.

68. A. Whitaker, *op. cit.*, p. 138.

69. S. Jones, *op. cit.*, pp. 194–195.

70. M. Jammer, *op. cit.*, p. 221.

71. M. Born and N. Wiener, *Z. Phys.* **36**, 174 (1926).

72. P. Ehrenfest to N. Bohr, 19 September 1925, in *The Emergence of Quantum Mechanics*, ed. by K. Stolzenburg, pp. 326–328; cited in M. J. Klein, in *The Lesson of Quantum Theory*, ed. J. de Boer, E. Dal, and O. Ulfbeck (North-Holland, Amsterdam, 1986), pp. 325–342.

73. S. Jones, *op. cit.*, pp. 168–180.

74. A. Schrödinger, The American Institute of Physics history program: https://www.aip.org/history-programs/niels-bohr-library/oral-histories/4865

75. M. Jammer, *op. cit.*, pp. 257–263.

76. S. Jones, *op. cit.*, p. 191.

77. J. Gribbon, *Erwin Schrödinger and the Quantum Revolution* (Wiley, New York, 2013).

78. E. Schrödinger, *Ann. Phys.* **79**, 361 (1926).

79. E. Schrödinger, *Ann. Phys.* **79**, 489 (1926).

80. E. Schrödinger, *Ann. Phys.* **79**, 734 (1926).

81. E. Schrödinger, *Ann. Phys.* **80**, 437 (1926).

82. L. Rosenfeld, in *The Physicists Conception of Nature*, ed. by J. Mehra (Riedel, Dordrecht, 1973), pp. 688–703, reprinted in *Selected Papers of Leon Rosenfeld*, ed. by R. S. Cohen and J. J. Stachel (Reidel, Dordrecht, 1979), p. 690.

83. D. Hilbert, *Math. Ann.* **59**, 161 (1904).

84. D. Hilbert, J. von Neumann, and L. Nordheim, *Math. Ann.* **98**, 1 (1928).

85. L. Rosenfeld, *Arch. Hist. Exact Sci.* **7**, 69 (1979), cited in J. A. Wheeler and W. H. Zurek, *Quantum Theory and Measurement* (Princeton Univ. Press, Princeton, 1983).

86. E. Madelung, *Z. Phys.* **40**, 322 (1926).

87. M. Born, *Z. Phys.* **37**, 863 (1926).

88. L. de Broglie, *Comp. Rend.* **184**, 273 (1927).

89. L. de Broglie, *J. Phys. Radium* **8**, 225 (1927).

90. M. Born, *Nature* **119**, 354 (1927); his notes were trans. by R. Oppenheimer for the publication; cited in M. Jammer, Ref. 8, p. 288.

91. A. Pais, *Niels Bohr's Times, in Physics, Philosophy, and Polity* (Clarendon Press, Oxford, 1991), p. 303.

92. P. McEvoy, *op. cit.*, p. 44.

93. W. Heisenberg, The American Institute of Physics history program: https://www.aip.org/history-programs/niels-bohr-library/oral-histories/4661-8

94. W. Heisenberg, *Z. Phys.* **43**, 172 (1927); trans. J. A. Wheeler and W. H. Zurek in *Quantum Theory of Measurement* (Princeton Univ. Press, Princeton, 1983), pp. 62–84.

95. S. Brush, *Graduate J.* **7**, 477 (1967).

96. A. de Gregorio, *Stud. Hist. Philos. Mod. Phys.* **45**, 72 (2014).

97. N. Bohr, *Nature* **121**, 580 (1928), reprinted in J. A. Wheeler and W. H. Zurek in *Quantum Theory of Measurement* (Princeton Univ. Press, Princeton, 1983), pp. 87–126.

98. G. Holton (1988), *op. cit.*, p. 134.

99. P. McEvoy, *op. cit.*, pp. 5–9.

100. P. McEvoy, *op. cit.*, pp. 70–8.

101. S. Jones, *op. cit.*, pp. 246–250.

102. M. Jammer, *op. cit.*, pp. 354ff.

103. J. Cushing, *Quantum Mechanics: Historical Contingency and the Copenhagen Hegemony* (Univ. Chicago Press, Chicago, 1994).

104. N. Levitt, *Phys. Today* **48**(11), 84 (1995).

Chapter 4

Solvay

Ernest Solvay was a Belgian chemist, who founded his own company and became a successful industrialist, having developed an important process for producing soda ash, a critical ingredient for making glass. He subsequently became a philanthropist, funding the Institute for Sociology at (what was then) the Free University of Brussels and The International Institutes for Physics and Chemistry. These latter entities coordinate and fund conferences, workshops, and seminars, mostly in Brussels. The Institute for Physics was founded in 1912 after the first "Conseil Solvay" in 1911. Since this first conference, there have been a series of Solvay Conferences, which continues today. These are usually focused upon an important (and usually argumentative) topic in either physics or chemistry. The most recent physics conference was the 2017 conference, the 27th in the series, which addressed questions in the physics of living matter. Perhaps the most famous one, however, was the fifth Solvay Conference, held in October 1927. Here, the announced topic would be electrons and photons, but the participants knew that the real meat of the conference would be the new quantum theories. But, as most participants also expected, the underlying current would be Einstein versus Bohr. Of the 29 invited attendees, 17 either were

The Copenhagen Conspiracy
David Ferry
Copyright © 2019 Pan Stanford Publishing Pte. Ltd.
ISBN 978-981-4774-75-8 (Hardcover), 978-1-351-20723-2 (eBook)
www.panstanford.com

already, or would become, Nobel prize winners, but this was quite common for the Solvay conferences.

The ideas that grew into the Solvay conferences arose from discussions between Walther Nernst and Max Planck. They became convinced that the problems that were currently raging in the theory of radiation and in the understanding of specific heat were sufficiently serious that an international meeting was required to bring all the scientists, with a vested interest in the topic, together to try and resolve these issues [1]. Nernst then had a chance encounter with Solvay, and this encounter provided the opportunity for the first conference to be held with Solvay's financial support. This would become the first international physics conference. Nernst then contacted Hendrik Lorentz and Martin Knudsen, a young Danish physicist. Lorentz would be the chair of the coming conference, and he put together a scientific committee which included Marie Curie, Marcel Brillouin, Emil Warburg, Heike Kamerlingh Onnes, Nernst, Ernest Rutherford, and Knudsen. Following the first conference, held at the end of October 1911, Solvay asked Lorentz about founding an institute to further these conferences. Lorentz' suggestions then led to the institutes mentioned above. Lorentz himself would chair the first five of the subsequent physics conferences. The first conference would be as famous for its scandal as for its science.

We consider the first conference, because it set the format for the majority of the subsequent conferences. The topic of this first conference was "the theory of radiation and quanta." Designed as a one week intense workshop, there would be roughly three days of talks by the invited participants, as well as some of the committee members. Then there would be two days of intense focused discussion. It was expected that this discussion would lead to some agreement among the participants upon the understanding of past experiments and theory and on the directions for future work. This did not always occur. Some of the conferences are remembered more for this discussion than for the talks themselves. This would be the case for the fifth conference in 1927. Nevertheless, the first and fifth conferences, in 1911 and 1927, have become legendary in physics.

Solvay 1911

It has to be remarked that there was no real tradition of scientific meetings at the turn of the twentieth century. In this regard, this first Solvay conference started as a novel kind of meeting [2]. It would be the first time that a significant number of physicists were brought together for an extended period of time. The letters of invitation to the first Solvay conference were dated June 15, 1911, and the conference dates were set as October 29 to November 4 of that year. The conference would be held at the Hotel Metropole. The actual conference would only be Monday to Friday, but the extra days allowed for socialization among the attendees. In addition to the conference committee, invitations went to Jean Baptiste Perrin (a French physicist who studied Brownian motion), Willy Wien, Henri Poincaré, Max Planck, Heinrich Rubens (of black-body radiation fame), Arnold Sommerfeld, Friedrich Hasenöhrl (an Austrian physicist who succeeded Boltzmann in Vienna, and was a friend of both Lorentz and Kamerlingh Onnes), James Jeans (a British physicist), Ernest Rutherford, Albert Einstein, and Paul Langevin. In addition, invitations went to Lord Rayleigh, J. D. van der Waals, Joseph Larmor, and Arthur Schuster (a German-born British physicist known for his spectroscopy efforts), but these four were unable to attend. Lord Rayleigh did send in a brief letter that became part of the proceedings. The absent were replaced by Georges Hostelet (a Belgian mathematician and philosopher) and Edouard Herzen (a Belgian chemist) who were scientific advisors to Solvay. As some record of the discussion and papers was necessary, three additional scientists were invited to serve as secretaries, whose task was to record the proceedings. These three were Robert Goldschmidt (a Belgian chemist), Frederick Lindemann (a British physicist, later Viscount Cherwell, who studied specific heats and confirmed some of Einstein's theories), and Maurice de Broglie (older brother of Louis, and 6th duc de Broglie).

The group of attendees was a mixture of what could be called the "old guard" and the "up and coming young stars." Planck and Einstein were the current leaders of these two groups. Einstein was the second youngest official attendee, being older than only Lindemann.

However, Maurice de Broglie brought his younger brother Louis along to observe these famous scientists. It is generally thought that this prompted the young Louis to devote his career to physics, and Maurice was likely particularly proud when Louis was invited as a speaker to the fifth conference. An interesting footnote to the conference is the requisite conference photo, in which Solvay was not present. A stand-in was found for him, and his headshot was pasted over that of the stand-in prior to the printing of the copies. Of course, as founder, his head was made larger than that of any of the participants. The official proceedings were published as a book (in French) and edited by Paul Langevin and Maurice de Broglie [3].[a]

The opening welcome and closing remarks were made by Solvay himself. Then, Lorentz set the stage with an opening talk about the problems in the area of radiation and quanta, where quanta was understood in the context of light quanta. In the proceedings are the manuscripts provided by Lorentz (in order of presentation), the letter from Lord Rayleigh, Jeans, Warburg, Rubens, Planck, Knudsen, Perrin, Nernst, Kammerlingh Onnes, Sommerfeld, Langevin, and Einstein. In addition, of course, the transcription of the discussions is provided, mainly following each of the talks. Perrin provided the longest manuscript and invoked perhaps the shortest discussion period, yet it was a good review of the experimental evidence for the existence of atoms. Planck and Einstein, of course, provoked significant discussion following their papers. While Planck's theory had been confirmed by Rubens and Einstein, among others, the relativity theory was still awaiting experimental confirmation. Instead, Einstein talked on the problems with specific heat, which connected to Planck's theory and the work of several other attendees. It is interesting that the manuscripts mentioned were actually prepared prior to the conference and distributed to the attendees in advance [4].

In his opening talk, Lorentz discussed the application of equipartition to the black-body radiation problem, therefore bringing into focus the other talks on black-body radiation. As mentioned, Planck's paper and talk invoked significant discussion. In fact, it was the main paper on light quanta. As such, it provides our link to the development of the later versions of quantum mechanics. Much of

---

[a]It is known that the proceedings were also published in German a year or two later.

the discussion of this paper was begun by Einstein, who had trouble with Planck's use of the Boltzmann equation and the statistical meaning of different configurations with the same probability [4]:

> The interesting part of the discussion centred on the question of whether, in our terminology, the electromagnetic field in vacuum ... could remain classical or had also to be quantized.... Planck expressed again his view that the quantum of action plays only a role in emission and absorption processes.

At this point, apparently Langevin pointed out a very recent paper by Debye, who directly quantized the field by considering that it also was to be considered as a collection of oscillators [5]. This problem would remain a thorn in quantum mechanics for at least the next several decades, and we return to it shortly.

The remaining talks seem not to have provoked much discussion or provided new looks into the future of physics. As Straumann [4] remarks,

> The scientific program ended with a general debate. This was opened with some brief remarks by Poincaré. He correctly states that in talks and discussions arguments were often based partly on the old mechanics, but in addition also on hypotheses which are in contradiction to it. Then he comments with a statement only a mathematician can make, and which I do not even find funny, namely that it is possible without great trouble to prove any statement, if the proof is based on two contradicting premises. On the bases of such an attitude, quantum mechanics would never have been discovered. The few other comments, in particular by Brillouin, Nernst, Poincaré, and partly by Langevin, are dominated by rather conservative hopes and statements.

Thus, in spite of the hope of the organizers in setting a new course for advancements in physics, the conference provided little in the way of drama or new ideas, although it did demonstrate one major point [6]:

> The new topic [of physics] was the quantum theory. It raised new, grave problems; for example, how could one understand probabilistic behavior on the atomic scale?

Yet, it would become famous for other reasons.

At the time of the conference, Marie Curie was just a few days short of being 44 years old, and it was just over five years since the death of Pierre. He had apparently, while in an absent-minded state of deep thought, wandered in front of a horse-drawn cart. They had been married just over a decade. But, at the start of the conference, the scandal of her affair with Paul Langevin became public. Paul's vengeful wife had gone to the newspapers with a story that he had eloped with Marie [7]. The resulting public outcry led to the scandal. In fact, she was also just announced as the winner of her second Nobel Prize (this time in chemistry), and the Swedish Academy went so far as to ask her not to attend the award. The scandal became an underlying theme for the Solvay conference, as many attendees were at odds as to how they should treat the couple—singly or as a pair. Einstein probably just proceeded normally, as his own belief was that she was not attractive enough to be a femme fatale [7]. For her part, Curie found an angry mob awaiting her upon her return to Paris, and had to seek refuge with friends for her daughters and herself. Langevin would fight a duel with a journalist who had vilified Curie, but it turned into a bloodless encounter. Time has a way of easing such things, and the scandal, as well as the romance, would eventually die away, although Curie eventually suffered a breakdown the following year. Nevertheless, it provided some discussion and adventure to the conference, perhaps more than the actual scientific discussion, if we are to believe the above comments.

The conference was also important as a coming-out event for the young Einstein. It was just six years since his main publications. He had just become full professor in Prague. As such, he was one of the youngest full professors in the Germanic world. To Einstein, it may have seemed akin to the trials of Prince Tamino in Mozart's *Der Zauberflöte*, trials which were part of the Prince's initiation into the body of priests at the temple of Isis and Osiris.[b] Along this same theme, Einstein himself referred to it as a "witches' Sabbath" [8]. Perhaps because of this, Einstein would emerge from the conference

---

[b]Isis is sometimes regarded as the patroness of nature, while Osiris is thought of as the god of regeneration and transition. Both are appropriate for one, such as Einstein, interested in learning the secrets of nature.

far more famous then when he arrived. Hence, the conference provided his transformation from one of the "up and comers" to perhaps the leading physicist in the Germanic countries, if not the entirety of Europe. It was his initiation into stardom.

At the end of the conference, Niels Bohr was visiting Manchester just after Rutherford returned from Brussels, and [9]

> was excited about having met Einstein and Planck. He regaled Bohr with stories of what had happened in Brussels. Bohr was so inspired by Rutherford that he decided to move to Manchester, keen to start working on Rutherford's new model of the atom, the one with electrons whirling around a nucleus like a miniature solar system.

It thus appears that the conference provided the inspiration for Bohr to begin work that would lead to his own breakthrough.

In retrospect, one wonders if the 1911 conference was really supposed to suppress the new light quantization. Just before the conference, Willy Wien was announced as the winner of the 1911 Nobel Prize in Physics (thus, two of the 1911 recipients were in attendance). Hence, Wien was being awarded for work that was known to be incorrect. Both experimentally and theoretically, it had been found that his formulation of the radiation spectrum of the black body just didn't fit the data. It is true that the Swedish Academy decided that his relation between the peak wavelength and the temperature was the important point, but it appears that they were hoping that new quantization would just go away. While there may have been some of this feeling at the conference, it was clear that light quantization was not going away. In fact, the youngsters Langevin (only a few years older than Einstein) and Einstein dominated the discussion sessions. The older Planck also contributed significantly. But it was clear that the younger generation was fully enveloped with the new approach and the quantum emphasis would only grow. It was only a year later that Bohr pushed the field forward dramatically. Yet, it would take almost another decade before the Swedish academy bothered to recognize Planck for starting this revolution. And the seeds of the second coming had been planted. Planck thought that Maxwell's fields did not need to be quantized, Einstein supposed that it would be necessary to do so, and Langevin mentioned the work of Peter Debye, who had already begun to do so. What Debye had done was to extend the theory by assuming that the vibration amplitude of the oscillators, or fields, contained a number of quanta, all with energy $\hbar\omega$ [5]. He would go beyond this by extending the ideas of harmonic oscillators to any arbitrary periodic motion in one dimension [10]. This was the beginning of quantum field theory. While Planck thought that the quantization was only important in emission and absorption processes, the later work of Compton would establish that the radiation field was quantized and that these photons carried momentum and energy [11]. Debye had the same idea [12], but would later say [13], "I published it, and then I found out that Compton had already published something. So during the time I was publishing it I found out that Compton had done it."

The world would be dramatically different sixteen years later at the fifth conference. Europe will have gone through the great war, famine, financial catastrophe, and a great many political changes leading to many new countries on the map. Einstein would be the older generation. Planck will have retired and been replaced by Schrödinger in Berlin. Heisenberg will have just been made full professor in Leipzig. He then will be the youngest full professor in Germany. Bohr will have joined the older generation stars, and a new generation of young "up and comers," which includes Heisenberg, will fill the conference with discussion and arguments.

Solvay 1927: Setting the Stage

Once more, the conference was to be planned by Lorentz, with the help of his committee. This was made more difficult by changes in the committee during the 1926–7 time frame. The final committee was Lorentz, Knudsen (who served as the secretary of the committee), Marie Curie, Albert Einstein (who replaced Kammerlingh Onnes in April 1926), Charles-Eugène Guye (a Swiss physicist who studied the velocity dependence of the electron mass), Paul Langevin (a French physicist who we met earlier), Owen W. Richardson (a British physicist who studied thermionic emission, and would win the Nobel prize in 1928), W. H. Bragg (another British physicist, who gave way to Blas Cabrera from Spain in May 1927, and had won the Nobel Prize with his son in 1915), and Edmund van Aubel (a Belgian metallurgist, who had nominated Kamerlingh Onnes and Nernst for the Nobel Prize in 1912—the former would win in 1913). The situation was more complicated at this time, since the Germanic scientists had been banned from most conferences in other parts of Europe since the war. Einstein had been considered as an exceptional case and invited to the 1924 conference, but had declined over the situation. The topic of the 1927 meeting, electrons and photons, would not provide a very good conference without the inclusion of the Germanic scientists, who were providing the most current theories. Already in early 1926, this problem was being discussed among the committee and Solvay. In fact, Lorentz had to travel to Belgium for an audience with the king to convince him to open the conference to the Germanic scientists. Bringing Einstein onto the committee was supposed to signal a receptive view to the Germanic scientists, but as late as October 1927, the issue was still sensitive. Van Aubel, even though a member of the committee, declined to attend. Nevertheless, the Germanic scientists had been to Como, Italy, earlier in the year. Whether Italy was as strict as Belgium is not known, but certainly Italy had not suffered the damage that Belgium and France had incurred in the western stage of the war.

Lorentz wanted Einstein to speak, and invited him to do so. Einstein originally planned to do so, but declined in June, and suggested that Schrödinger replace him. As we will see below, Einstein had been working on a new wave mechanics approach, but abandoned it and asked for Schrödinger to replace him. Schrödinger had not been on the original list of proposed speakers, probably because his work on the wave equation was still very new, but Planck had seen the manuscripts and called Einstein's attention to the coming work. The idea was accepted, and the final program would include talks by W. L. Bragg, A. H. Compton, de Broglie, Born and W. Heisenberg together, and Schrödinger. Each talk was assigned to a half-day session, with the talk to be followed by intense discussion. A problem had arisen in the scheduling of the conference, since it was discovered that the planned dates overlapped an important event in Paris—the celebration of the centenary of Fresnel. So, it was decided to allow the attendees to travel on Thursday to Paris for this event, returning to Brussels on Friday morning. Then the general discussion would take place Friday afternoon and Saturday morning. The first two talks would focus on X-ray scattering and the last three would present the new theories of quantum mechanics.

From the committee, Lorentz, Curie, Einstein, Langevin, Guye, Knudsen, and Richardson would attend, as would Solvay's science advisor Emile Henriot from the Free University of Brussels. In addition to the six speakers, the other attendees would be Auguste Piccard (a Belgian physicist), Ehrenfest, Herzen, Théophile de Donder (another Belgian physicist), Pauli, Fowler, Leon Brillouin (son of Marcel), Debye, Kramers, Dirac, Bohr, Irving Langmuir (an American chemist), Planck, and Charles Wilson (the British inventor of the cloud chamber). In addition, Jules Emile Verschaffelt would be the general secretary for the conference, and would attempt to record the discussion. Of course, the speakers were asked to provide their manuscripts, which would be distributed to the various attendees in advance of the conference.

The conference would occur when there was significant tension in the air. Born had essentially broken off communicating with Einstein.[c] Schrödinger had gotten Planck's position in Berlin, and it was known that Born was also a candidate for the position, whether or not he wanted to leave Göttingen. Schrödinger had gone to the United States in January 1926 for an extended visit to various universities, and probably wasn't overly happy since this was the middle of the prohibition era. Schrödinger and Lorentz met in Pasadena, and this helped Lorentz with his understanding of the wave theory, and probably helped to cement his later selection as a speaker at Solvay. But his absence left no one to really drive wave mechanics forward, and the field was open for the Göttingen crowd to push their theories forward. The new paper on uncertainty and Bohr's paper on complementarity certainly drove this field into new directions. During the year, de Broglie was in the process of expanding his theory of quantum mechanics, with the double

---

[c]Even in Born's own book dealing with his correspondence with Einstein, there is a significant gap during this period.

wave conjecture [14]. In this viewpoint, every solution for the wave function had a corresponding singularity solution having the same phase of the wave function [15]. In addition, the velocity of the singularity could be shown to be given by the gradient of the phase of the wave function, and hence corresponded to the particle motion. This meant that it was the phase of the wave function, rather than the amplitude, that determined the motion of the particle that was represented by the wave function. Thus, he had reconciled the particle and wave picture in his mind [16]. In this picture both the wave and the particle were real things, and this differed from either Heisenberg or Schrödinger. In a sense, this was a third approach to quantum mechanics. Moreover, the only probability is in the initial conditions for the wave function. As John Bell [17] would put it, the de Broglie paper

> is perfectly conclusive as a counter example to the idea that vagueness, subjectivity, or indeterminism, are forced upon us by the experimental facts covered by nonrelativistic quantum mechanics.

In addition, Einstein was to weigh in as well. In May 1927, he gave a talk to the Prussian Academy of Sciences on the wave mechanics viewpoints. While he withdrew his manuscript from the proceedings, copies still exist in his archives [18]. (This is the lost manuscript referred to by Born in the last chapter.) Yet, the topic paper was known to the attendees in Berlin, and has been discussed widely. A brief overview appears in Ref. [15], and the paper is seriously reviewed by Belousek [19]. The opening statements of the paper are [19]:

> As is well known, the view currently prevails that a complete space-time description of the motion of a mechanical system does not exist in the quantum-mechanical sense. For example, it is supposed to make no sense to speak about the instantaneous configuration and velocities of the electrons of an atom. In contrast, it shall be shown in the following that Schrödinger's wave mechanics suggests that one assign motions of the system uniquely to each solution of the wave equation.

He considers the problem in configuration space (a space of $3N$ coordinates, where N is the number of particles) and shows that a proper solution can be obtained which will describe the positions of the particles. In this approach, the velocities of the particles would be considered as "hidden" variables, since they were not directly observable. This would then be the first of many so-called hidden-variable approaches to quantum mechanics. Why Einstein withdrew the paper is, of course, unknown. It is likely that he found failures within his own approach that made the results untenable. Cushing [15] suggests that Einstein considered non-interacting particles, whereas the real multiple particle picture would have to consider the presence of interactions. This would break up his principle approach to the flow lines (trajectories) which Einstein used, and this was untenable. In a sense, the interaction among the particles would produce an "entangled" state of the system. This failure by Einstein likely colored his view of wave mechanics for some time, as would be apparent in the conference.

But the idea of hidden variables was not very new. Theodor Kaluza pointed out to Einstein, already in 1919, that if he solved the equations for relativity in five dimensions (four spatial dimensions plus time), Maxwell's equations for electromagnetics fell out of the picture naturally. With Einstein's encouragement, he published the paper in 1921 [20]. As a result, Einstein, Ehrenfest, and Klein were all known to be working on higher dimensional spaces in 1926–7 [21–23]. As Jones says, "They weren't the only physicists who recognized that, theoretically speaking, Bohr's quantized atom in three spatial dimensions appeared to become a classical atom in four" [24].That is, quantization appeared only when the fourth spatial dimension was taken out of the picture. It was Klein, however, who pointed out that the fourth spatial dimension probably was not observable because (in his view) it would be rolled up into a series of very tiny circles, whose dimension was less than the Planck length (a length of about 10^{-35} m). Hence, this fourth dimension was *hidden* from any observer in three spatial dimensions.

In addition, in classical statistical physics, *it had always been recognized that the need for a statistical approach lay in the impossibility of actually being able to describe all the variables necessary* to solve for motion of the very large number of particles that existed in, e.g., Brownian motion. Thus, the statistics describes the inability to know all the variables. Then, it would seem that if quantum mechanics was to be a statistical theory, as Born had said, then there should be some hidden dimensions, or variables. In fact, McEvoy [25] states it thus: "It is up to those advocating the statistical interpretation to show why the concepts used in the description ... have only a statistical significance; and they must show that the embedding of quantum mechanics in a more detailed hidden variable theory is mathematically possible." Thus, it seems that the probabilistic view would require some hidden variables, and it would be incumbent upon Göttingen to show what they were, rather than trying to crush anyone's suggestions as to their possible existence. Yet, that would not happen.

And Schrödinger was still complaining about quantum jumps. Initially introduced with Bohr's atomic model to explain how the electron got from one state to the other, it had remained an essential part of the Göttingen matrix mechanics. However, in the wave mechanics approach, perturbation theory allowed the wave to move smoothly from one state to the other. There was no necessity for a quantum jump. This smooth transition led many to consider the wave mechanics approach the far superior approach. Rosenfeld [26] put it thus:

> The similarity of structure between particle mechanics and geometrical optics, pointed out long ago by Hamilton, suggested that the wave aspect be the fundamental one, whereas the particle trajectories would correspond to an approximate description of the same type as the rays of light.

To Schrödinger, the matrix approach had no underlying theoretical basis. The matrix elements connecting different states were an *ad hoc* concept. Consequently, their existence required no dependence on causality, since there was nothing to induce the transition. Such an interpretation depended upon Bohr's philosophical approach in which the measurement produced any needed reality. On the other hand, both Schrödinger's wave mechanics, and de Broglie's wave–particle picture, provided an underlying theory and re-established a causal connection between a driving quantity and the resulting transition matrix elements.

Then, there was certain to be friction between Einstein and Bohr. Their dispute had become known to the press years before. While the press attention had died down, it was known in scientific circles that the dispute was alive and well. The attendees would certainly also know that Ehrenfest had brought them together in Leiden in an attempt to have them resolve their differences. It had also failed. For sure, there would be extensive discussions among the participants, and well known that there would be give and take between Einstein and Bohr. As Bacciagaluppi and Valentini [1] remark, "Informal discussions at the conference must have been plentiful, but information about them has to be garnered from other sources." Primarily, this seems to be due to the fact that the Einstein–Bohr interaction largely took place outside the general discussions that followed the presentations and in the Friday–Saturday sessions. As Heisenberg [27] later recalled,

> The real fight was between Einstein and Bohr. Probably you have been told many times of this meeting. Usually, in the morning at breakfast, Einstein would have invented a new experiment by which he definitely could disprove the theory, and he would say, "Well, there you can disprove it. Certainly it doesn't work." Then Bohr would be in despair and would discuss the thing, and by the evening, again, Bohr would be on the winning side. He would say, "Well, this is the interpretation and there you see it can work." Einstein would be in despair and the next morning he would come back with a new example.
>
> The Brussels meeting was the one where really the details were discussed and where one had to fight. There Einstein tried very hard to disprove the theory. He was so dissatisfied that he really insisted on the fact that there must be something wrong and he was always very angry when he found that he had not come through with it. That was a real hard fight which started at breakfast in the morning. I used to follow Einstein and Bohr on our way from the breakfast to the meeting, which was about five hundred yards down the street. They would talk to each other in a very energetic way, discussing it all the way. At the breakfast, usually, Einstein was on the winning side because then he had a

new problem and Bohr would try to think about it. Then during the lunch time, of course, Pauli and I discussed [the] new problems. In some cases, the answer was simple but there were a few cases like this light quantum that the answers were far from being trivial, but were rather deep.

... the problem would be discussed very eagerly among Bohr and Pauli and myself. I think it was always the three of us. But then Bohr would remain alone in his room for a few hours and Pauli and I would discuss it and in the evening we would come together again and say, "Well, what do you think now about it." But mostly I would say it was Bohr who was very eager to give the correct answer by at least late in the evening.

It is very clear from these statements that the Einstein–Bohr debate was really an undercurrent within the meeting, but that most of the attendees were outside this debate. It would certainly explain why Verschaffelt had not included these issues in the proceedings—he was likely unaware of them, except perhaps superficially. But watching the debate was Paul Ehrenfest. As Heisenberg [27] remarked,

Ehrenfest finally said, "Einstein, ich schame mich fur Sie." "I am ashamed for you because these discussions are just like those about relativity, and now I see that Bohr is right and you don't believe it."

Heisenberg had great respect for Ehrenfest, and, in some sense, the latter was the conscience of the quantum mechanics community. He certainly tried to moderate the Einstein–Bohr dispute, but he would always be there to ask critical questions. In response to a question about whether it was a problem having Ehrenfest around, Heisenberg [27] commented,

Yes, to some extent. As soon as somebody came with an idea which was a bit constructive ... he would be a bit bothered by a man who would always say, "It's just wrong." On the other hand, everybody agreed that the situation was very bad, but since nobody could offer a solution it didn't bother people so much as it could perhaps have. Obviously, also, Ehrenfest couldn't do any better. Ehrenfest in that way was much better than, say, people like Wien, who simply would throw away the whole quantum theoretical business and would say, "Well, that's all full of contradictions, therefore, forget about it." Ehrenfest would just never forget about it. He would say, "We have to think day and night about it just because it's so disagreeable."

... Uhlenbeck also said, "I think Born was also a little bit afraid of Ehrenfest," because ... Born could put it into mathematics and then Ehrenfest would look at it and say, "Yes, but what about the physics right there?"

... he didn't like too much the purely phenomenological way of arguing, and he didn't like Sommerfeld's mathematical schemes and Born's more complicated mathematics. Yet, I think that was a type of physics which was absolutely necessary at that time, otherwise the thing never would have come out. I think nowadays it is done too little. We have really too few Ehrenfests now.

Some have remarked that Einstein was trying to disprove quantum mechanics in the experiments he would propose to Bohr, but this is unlikely. He had brought quantum effects to the forefront in his 1905 paper, and he had tried to develop his own approach to wave mechanics earlier in the year, although he felt he had failed and never published the paper. It is more likely that his objection was really to Bohr's philosophical concepts, which denied reality, other than in the measurement, and determinism. As Pais [28] would say,

It was his almost solitary conviction that quantum mechanics is logically consistent, but that it is an incomplete manifestation of an underlying theory in which an objectively real description is possible—a position he maintained until his death.

From this, it is clear that the differences between Einstein and Bohr would never be reconciled. The debate would continue for years.

Another event, that would color the discussion, was quantization of both the electromagnetic field and Schrödinger's wave equation. As remarked above, Debye already started the process of quantizing the electromagnetic fields in 1911. Dirac also quantized these fields in an important paper in mid-1927 [29]. But then, in a new paper just before the conference, he noted that the new

Schrödinger equation was continuous and had trouble producing discrete emission and absorption processes, at least in his view of the situation. So, he proceeded to show that he could also quantize the wave equation as a series of oscillators, much like those in the electromagnetic field quantization [30]. As Paul McEvoy [31] put it,

> By analogy to the theory of the electromagnetic field, the particle wave function was soon quantized, given its own commutation relations, and renamed a field function. This procedure was called 'second quantization' at the time.

We have discussed fields and vector fields earlier, but this approach now mathematically equated the wave function in space to the same status as one of the electromagnetic fields, although not a vector field. The advances to quantum field theory had taken another major step.

As if the above problems weren't enough, the stage was to be set by the opening two talks, to be given by Bragg and Compton. It was to be continuum versus discontinuity. Bragg was to talk on the use of X-rays to study materials via Maxwell's electromagnetic wave approach. Compton was to talk about X-ray scattering which involved the quantum nature of light. So, the committee had clearly decided to set the stage to provide the problem and then the new quantum mechanics that should better describe the situation (or the problem, depending upon your viewpoint). As Compton would say in his introduction [1],

> Professor W. L. Bragg has just discussed a whole series of radiation phenomena in which the electromagnetic theory is confirmed.... I have been left the task of pleading the opposite cause to that of the electromagnetic theory of radiation, seen from the experimental viewpoint.

Solvay 1927: The Conference Itself

While Lorentz chaired the conference and may have given an introductory, and welcome talk, no manuscript exists of this. Lorentz died shortly after the meeting, and his comments were replaced by an obituary written by Marie Curie. Hendrik Lorentz was a well-respected person and scientist, a point made abundantly clear by Curie [1]:

> All those who had the honour to be his collaborators know what he was as chairman of these conferences and of the preparatory meetings. His thorough knowledge of physics gave him an overall view of the problems to be examined. His clear judgement, his fair and benevolent spirit guided the scientific committee in the choice of the assistance it was appropriate to call upon. When we were then gathered together at a conference, one could only admire without reservations the mastery with which he conducted the chairmanship.

Lorentz would be replaced by Paul Langevin as chairman of the committee.

As mentioned above, the first lecture of the conference was given by William L. Bragg, the Australian born physicist famous for his work in X-ray studies of crystallography. He discussed the classical theory of X-ray scattering and its use in studying the arrangement of atoms in crystals. Here, the theory is based not upon scattering from individual atoms, but from planes of atoms. The younger Bragg came to the field following the work of his father W. H. Bragg. Bragg discussed the historical development of the theory and of the experiments, and also about the interpretations that could be placed upon the various results. He then discussed how these measurements connected to the current concepts of quantum mechanics. For example, following the talk, Debye asked how he could conclude that the crystal had an energy even at absolute zero temperature (Debye himself had contributed to the theory of scattering). Bragg responded, in part, as [1]:

> If one assumes the existence of such an energy, the curve deduced from the experimental results agrees with that calculated by Hartree by applying Schrödinger's mechanics. On the other hand, the curve that one obtains if one does not assume any energy at absolute zero deviates considerably from the calculated curve by an amount that exceeds the possible experimental error.

One may suspect that Debye knew the answer before he asked the question, especially since the manuscripts had been distributed prior to the meeting, and because he was himself a competent scientist (he would get the Nobel Prize in Chemistry in 1936 for his work on dipole moments of molecules). But he wanted to make an important point and to insure that it went on the record. The zero point energy is the remaining energy that the oscillators (here the description of the atomic motion) have even at absolute zero was important. We learned in the last chapter that the energy of the oscillator was $(n + 1/2)\hbar\omega$, so even when the system was at absolute zero, and the atoms were all in their ground state $(n = 0)$, the atoms still moved with an energy that was one-half the oscillator energy quantum. This result was common to both the Heisenberg and the Schrödinger formulations of quantum mechanics, but it did not appear in the Planck and Bohr view of the oscillator energy. So, this so-called zero-point energy was unique to the new quantum mechanics. The point that Debye was making was that even with the classical wave approach, the zero point energy of the atomic motion clearly shows up as a quantum effect, agreeing with the modern view of the energy levels of the various oscillators. And it was necessary to include this quantum effect in order to correctly calculate the magnitude of the scattered X-rays. Hence, even with Bragg's wave picture of light, quantum mechanics got involved in determining the strength of the scattered X-rays. This was a significant point. There followed a general discussion, to which several of the attendees participated, on the theoretical basis for the quantum scattering that had been carried out by Hartree. It is clear that, since Bragg was the first speaker, many of the younger attendees were striving to demonstrate their abilities by asking questions in this first session, perhaps staking out a claim for their own relevance. Einstein and Bohr said nothing.

Then came Compton. He would present the X-ray scattering theory which led to what has become known as Compton scattering. In the translation from the French official report, he comments [1],

> I have no intention of diminishing the importance of the electromagnetic theory as applied to a great variety of problems. It is, however, only by acquainting ourselves with the real or

apparent failures of this powerful theory that we can hope to develop a more complete theory of radiation.

Thus, he discusses basically five points which are problems that must be considered. These were: (i) is there an ether, or medium supporting the waves; (ii) how are the waves produced; (iii) the photoelectric effect; (iv) the scattering of X-rays and the recoil of the electrons (Compton scattering); and (v) experiments on "individual interactions between quanta of radiation and electrons." He uses the term "photon" and credits its source to Lewis [32]. Of course, he spends considerable time discussing the scattering problem because [1]

It is impossible to account for scattered X-rays of altered frequency, and for the existence of the recoil electrons, if we assume that the X-rays consist of electromagnetic waves in the usual sense.

Of course, the scattering was shown to arise as the photon lost energy to the electron, which in turn changed the energy of the outgoing X-ray. He makes a point in the talk of discussing the BKS paper (Chapter 2) and comparing the theoretical results with the discussion of virtual oscillators introduced in that paper. He does comment [1],

It is difficult to see how such a theory could by itself predict the change of wavelength and the motion of the recoil electrons. These phenomena are directly predictable if the conservation of energy and momentum are assumed to apply in the individual actions of radiation on electrons; but this is precisely where the virtual radiation theory denies the validity of the conservation principles.

He finishes by a thorough discussion of a wide range of experiments, both on single particle scattering of X-rays and of various versions of the photoelectric effect. In the discussion following the talk, there was much comment on the various experiments. Bohr felt motivated to make a somewhat confusing comment about the need for four wave trains (two for the electron and two for the X-ray, one each before and after the interaction), and since these were assumed to be nearly single frequency, they would be fairly extended in space. He worried about the space-time considerations and whether they were being violated. A more interesting comment was made by Lorentz, who clearly understood the questions raised by the experiments of Taylor years earlier (mentioned in Chapter 1) [33], and said [1],

> It is quite certain that, in the phenomena of light, there must yet be something other than the photons. For instance in a diffraction experiment performed with very weak light [Taylor's experiment, for example] ... there are instants when no photon is present in the space under consideration. This clearly shows that the diffraction phenomena cannot be produced by some novel action among the photons. There must be something that guides them in their progress and it is natural to seek this something in the electromagnetic field, with its waves.

Compton responded that indeed it was hard to avoid the waves. And now we have the reassertion of the idea of particle *and* wave being necessary to explain experiments. After further discussion on the momentum of the electron and how it is related to that of the X-ray, Curie discussed how the collision viewpoint could apply to biological tissue. Ehrenfest induced Schrödinger to draw the interaction of the collision and the four waves that could be described in his wave picture, in what was described as an *anschaulich* (descriptive) fashion. Then Bohr felt obliged to comment [1]:

> The simultaneous consideration of two systems of waves has not the aim of giving a causal theory in the classical sense, but one can show that it leads to a symbolic analogy.... it has been possible to treat the problem in more depth through the way Dirac has formulated Schrödinger's theory. We find here an even more advanced renunciation of Anschaulichkeit [intuitive].

It is not clear here whether Bohr is trying to deny causality or say it appears in quantum theory by analogy, but he also apparently feels he has to take a jab at Schrödinger. Again, Lorentz remarks that Schrödinger succeeded in his task, at least to his satisfaction, perhaps his own renunciation of Bohr's philosophy.

The third lecture was that of Louis de Broglie. It is generally felt that de Broglie backed away from his 1927 paper on the double wave theory. He gave a general overview of both matrix mechanics and wave mechanics, and a long discussion on the concept of electrons as waves. He discussed several experiments which seemed to confirm the wave nature of the electrons. In the discussion period, Born raised the question [1]:

> The definition of the trajectory of a particle that Mr. de Broglie has given seems to me to represent difficulties in the case of a collision between an electron and an atom. In an elastic collision, the speed of the particle must be the same after the collision as before. I would like to ask Mr. de Broglie if that follows from his formula.

De Broglie assured the group that it did follow from his theory. The success was actually then pointed out by Brillouin [1]:

> It seems to me that no serious objection can be made to the point of view of L. de Broglie. Mr. Born can doubt the real existence of the trajectories calculated by L. de Broglie, and assert that one will never be able to observe them, but he cannot prove to us that these trajectories do not exist.

Many have suggested that Pauli energetically opposed de Broglie's ideas and crushed the idea. This isn't supported by the proceedings. Following de Broglie's talk, Pauli made the comment [1]:

> It is always possible to introduce a velocity vector ... and to imagine furthermore corpuscles that move following the current lines of this vector. Mr. de Broglie now introduces an analogous representation for material particles. In any case, I do not believe that this representation may be developed in a satisfactory manner; I intend to return to this during the general discussion.

And he did, and this will be discussed below. While the overall discussion was lengthy, it appears to have been carried out in a respectful manner.

The next lecture was the combined talk of Born and Heisenberg. The talk is notable for the claim that they make [1]:

Quantum mechanics tries to introduce the new concepts through a precise analysis of what is "observable in principle." In fact, this does not mean setting up the principle that a sharp division between "observable" and "unobservable" quantities is possible and necessary.... At the same time the new conceptual scheme provides the anschaulich[d] content of the new theory.... Quantum mechanics is meant as a theory that is in this sense anschaulich and complete for the micromechanical processes.

Here we have it. This represents the first claim that the new quantum mechanics is complete and need not ever be modified. After reiterating all the mathematics that has previously gone into matrix mechanics, they discuss the implications. Once more, they assert [1],

One sees that quantum mechanics yields mean values correctly, but cannot predict the occurrence of an individual event. Thus, determinism, held so far to be the foundation of the exact natural sciences, appears to go no longer unchallenged.

They admit that statistical physics actually exists, where determinism has never been challenged, they offer no proof for its disappearance in the quantum world; they merely assert that this is the fact. The interesting part of the proceedings is that there was so little discussion after this talk. Primarily, Dirac talked about a connection between matrix mechanics and classical mechanics, and Lorentz seemed surprised that there was an equation of motion for the matrix representations. It is almost as if the assembly had become resigned to matrix mechanics as the new gospel, and its arrival from on high was not subject to question or discussion. Unfortunately, this concept would prevail for a great many years.

The final lecture was given by Schrödinger: "Wave Mechanics." In his introduction, he carefully distinguishes between de Broglie's approach and his own. For sure, it is important to him to make sure that his work is not seen as merely an extension of that of de

---

[d]This seems to be a favorite word, which is usually not translated, but means vivid or descriptive. It is not clear whether the translation makes the point as emphatically as the original does in German.

Broglie, as such a view would diminish his impact. Hence, he opens with [1]:

> Under this name [Wave Mechanics] at present two theories are being carried on, which are indeed closely related but not identical. The first, which follows on directly from the famous doctoral thesis by L. de Broglie, concerns waves in three-dimensional space. Because of the strictly relativistic treatment that is adopted in this version from the outset, we shall refer to it as the *four-dimensional* wave mechanics. The other theory is more remote from Mr. de Broglie's original ideas, insofar as it is based on a wave-like process in the space of *position coordinates* (*q*-space) of an arbitrary mechanical system. We shall therefore call it the *multi-dimensional* wave mechanics.

He then goes on to review his multidimensional theory, which includes separate coordinates for each of the particles that exist in the formulation. That is, he expands the wave function into the position coordinates of configuration space—$3N$ coordinates, where N is the number of particles. His development is carried out by the definition of a Lagrangian, rather than simple use of the Hamiltonian. He then discusses the reduction to the four-dimensional theory of de Broglie and the many-electron problem, making connection to matrix mechanics along the way. Bohr asked about the "difficulties" Schrödinger mentioned in regard to the intensity of light. Schrödinger answers that the problem is that in calculating these intensities using an expansion of the wave function into a proper basis set, one arrives at the intensity being proportional to the product of the square magnitudes of both the initial and final states, whereas the "old theory" predicts that only the initial state should appear. Bohr then asks whether Dirac has found the answer, to which Schrödinger responds [1],

> Dirac's results are certainly very interesting and point the way toward a solution, if they do not contain it already. Only we should first come to an understanding in physical terms ... I find it still impossible, for the time being, to see an answer to a physical question in the assertion that certain quantities obey

a noncommutative algebra, especially when these quantities are meant to represent numbers of atoms.

Born then enters the fray with a comment that the assumption that the squared magnitude of the wave function that Schrödinger uses leads to difficulties in the case of quadrupole moments. To which Schrödinger responds [1],

> I can assure you that the calculation of the dipole moments is perfectly correct and rigorous and that this objection by Mr. Born is unfounded. Does the agreement between wave mechanics and matrix mechanics extend to the possible radiation of a quadrupole? That is a question I have not examined. Besides, we do not possess observations on this point that could allow us to use a possible disagreement between the two approaches to decide between them.

Here, he is basically saying to Born, "Don't be foolish." There weren't such experimental data available, and Born was just being combative. Then follows a discussion between Fowler and de Donder about the many-electron problem. Born then shifts to discussing their own efforts to calculate many matrix elements with the purpose of providing a table of them, to which Lorentz asks whether such a task is really useful. Finally, Schrödinger, Heisenberg, and Born engage in a discussion over whether or not the approximations which Schrödinger had made to the many-electron problem was either satisfactory or even possible. This question is still being debated today.

From Schrodinger's lecture on Wednesday to the start of the general discussion on Friday afternoon, the attendees traveled to Paris, as mentioned earlier. This gave them a full day and a half to discuss the various lectures and to ponder upon the meaning and the problems. Thus, there was significant time allotted to the participants to plan on their approach in the general discussion. Apparently, Bohr did not use this time wisely, or perhaps he was required to use all his time in finding solutions to the problems Einstein would present each morning. The result was that most of Bohr's comments have been redacted from the proceedings and replaced by a copy of his Como paper. We will remark upon that later, but his comments must have been very obfuscating [1]:

> The extant notes relating to Bohr's remarks at this point are particularly fragmentary. Kalckar's introduction to volume 6 of Bohr's Collected Works [34] describes the corresponding part of notes (taken by Kramers and by Verschafflet) in the Bohr archives as too incomplete to warrant reproduction.

However, notes taken by Richardson, one of the committee members, were a little better and will be discussed below at the appropriate point.

The formal discussion session opened with comments from Lorentz, who had remarkable insight into not only the new theories, but the questions that these theories raised. He began by discussing the old (classical) theories where particles had clear trajectories, even when passing through the cloud of the atom. That is, he viewed the electron clearly as a corpuscle which existed at a definite point in space at a given time. Then, he came to the new theory [1]:

> I imagine that, in the new theory, one still has electrons. It is of course possible that in the new theory, once it is well-developed, one will have to suppose that the electrons undergo transformations. I happily concede that the electron may dissolve into a cloud. But then I would try to discover on which occasion this transformation occurs. If one wished to forbid me such an enquiry by invoking a principle, that would trouble me very much. It seems to me that one may always hope one will be able to do later that which we cannot do at the moment. . . . Let us take the case of an electron that encounters an atom; let us suppose that the electron leaves the atom and that at the same time there is an emission of a light quantum. One must consider, in the first place, the systems of waves that correspond to the electron and to the atom before the collision. After the collision, we will have new systems of waves. . . . I think I would find that, to explain the phenomena, it suffices to assume that the expression $\psi \psi^*$ gives the probability that the electron and the photons exist in a given volume; that would suffice to explain the experiments. But the examples given by Mr. Heisenberg teach me that I will have thus

attained everything that experiment allows me to attain. However, I think that this notion of probability should be placed at the end, and as a conclusion, of theoretical considerations, and not as an a priori axiom, though I may well admit that this indeterminacy corresponds to experimental possibilities.... Could one not keep determinism by making it an object of belief? Must one necessarily elevate indeterminism to a principle?

Here we see the essence of the arguments that will remain for decades about the interpretation of quantum mechanics. Where in the theory of Heisenberg and Born is the absolute requirement for indeterminism? To Lorentz, and others, there was no such requirement to eliminate determinism nor for eliminating causality. Moreover, he clearly points out that, at least to him, the current state of quantum mechanics, whether it be matrix mechanics or wave mechanics, was certainly not at a complete and satisfactory level of development.

Following Lorentz, Bohr made his comments, which do not appear in the transcripts other than in the fragmentary form discussed above. From Richardson's notes, we know that Bohr talked about the experiments which Heisenberg had discussed in his first paper on matrix mechanics. He also discussed Compton's experiments. Then, apparently, Richardson's notes jump to a section from the Como lecture where Bohr claims [1],

> In any observation that distinguishes between different stationary states one has to disregard the past history of the atom.

Paradoxically, one then assigns a phase to the stationary state (which must be spatially independent if it is to remain a stationary state). Yet, Bohr wants to have a partially open system and use a wave packet which has no well-defined phase. The fact that the notes are so fragmentary must imply that Bohr's discussion was difficult to follow, perhaps because he used his normal obfuscation approach. However, Brillouin pointed out that the uncertainty of simultaneous measurements of position and momentum was an old idea relating to the concept of cells in phase space that had been introduced by Planck. He credited Bohr for making the idea less abstract.

Born then discussed a problem proposed by Einstein, concerning α-particles. Since Einstein had not yet contributed to the general discussion, it is likely that this was one of Einstein's morning problems presented to Bohr, but one that was more known to the general audience. This is especially likely as Born refers to comments by Pauli during his discussion. It is not clear whether Bohr had obtained an answer to the question. Born expressed the problem as [1]:

> Mr. Einstein has considered the following problem: A radioactive sample emits α-particles in all directions; these are made visible by the method of the Wilson cloud [chamber]. Now, if one associates a spherical wave with each emission process, how can one understand that the track of each α particle appears as a (very nearly) straight line? In other words: how can the corpuscular character of the phenomenon be reconciled here with the representation by waves?

Born then points out that once ionization has occurred in the cloud chamber the spherical wave has collapsed due to the wave packet reduction introduced by Heisenberg. This would cause the wave to reduce to a series of drops along a trajectory or ray. He points out that Pauli wondered if this could be treated in a multidimensional space, which Born agrees is possible. He goes on to discuss this approach.

Then comes Einstein, who points out that one can take two possible positions toward the theory of quantum mechanics, and proposes a descriptive example. This is composed of a hemispherical structure with a screen stretched across the open side of the hemisphere. Photographic film is placed inside the hemispherical shell, and a small hole is placed in the screen at the center. Now, electrons are allowed to impinge upon the screen, some of which will actually pass through the small hole. Because the hole is small, the electrons which pass through it are dispersed uniformly (by diffraction) over the direction of the hemisphere; that is, they would uniformly strike the photographic film in all directions. Now, the quantum theory tells us that there are de Broglie waves which impinge upon the screen and are diffracted through the hole. Behind the screen, there are thus spherical waves. When these waves reach the screen, the intensity at an arbitrary point P determines whatever happens at this point. Now, in the first view, the de Broglie-Schrödinger waves do not correspond to a single electron, but to a cloud of electrons, so that the theory gives no information about individual processes. In the second view, however, the theory gives a complete view of individual processes, so that each particle is described by a wave packet, according to its position and speed, and this wave packet is diffracted through the hole. Now, comes the problem. In the first view, $\psi\psi^*$ gives the probability that there exists a particular particle from the cloud at point P. However, in the second view, $\psi\psi^*$ gives the probability that the *same* particle appears at P. But this latter view says that the theory refers to an individual process, yet this would allow the particle to arrive at different points on the screen with equal probabilities. Yet, only the second view allows for the conservation laws to remain valid. Einstein then goes on to say that, in his view, one must go beyond the Schrödinger equation to solve this dilemma, and gives support to the de Broglie approach for solving this problem. If one measures one of the points to be occupied, then the others must disappear through wave function collapse very rapidly, which may violate the speed of light limit. Now, many regard this problem as the opening salvo in what would become the EPR problem (Chapter 6). For his part, Bohr could only comment [1], "I don't understand what precisely is the point which Einstein wants to [make]." Bohr then tries to respond, but drifts off into a discussion of space-time concepts, and finally remarks that there are features that are complementary to the wave description and perhaps are important to this problem.

Lorentz suggests treating Einstein's problem in several spatial dimensions in configuration space, but Pauli points out that this would only be a simple technical means for handling the interactions between several particles, interactions which do not allow themselves to be handled simply in normal space and time. He then mentions approaches taken by Dirac and by Jordan and Klein which may provide another approach but essentially reduce to Schrödinger's multidimensional wave equation. He presents an example of two hydrogen atoms, each in their ground state, but separated by a great distance. He points out that the induced polarization effect of one atom on the other (and vice versa) yields a contribution to the total energy that is the largest correction.

Dirac then discusses further the experiments which can give difficulty when limiting oneself to only the normal three-dimensional theory. He then points out the present situation in multidimensions leads to the necessity of ignoring relativity theory, and then shows that Schrödinger's expression for the electric density appears naturally in the matrix theory. Dirac then turns to determinism in response to Bohr's earlier remarks. Here, he points out that [1]

> in the classical theory one starts from certain numbers describing completely the initial state of the system, and deduces other numbers that describe completely the final state. This deterministic theory applies only to an isolated system. But, as Professor Bohr has pointed out, an isolated system is by definition unobservable. One can observe the system only by disturbing it and observing its reaction to the disturbance. Now, since physics is concerned only with observable quantities the deterministic classical theory is untenable.

Now, it appears that this denies causality and determinism even in the classical world, but he is setting up a discussion of this in the quantum case. Here, he starts with the experimental decision following Bohr's eternal argument [1]:

> The disturbance that an experimenter applies to a system to observe it is directly under his control ... *it is only the numbers that describe these acts of freewill that can be taken as initial numbers for a calculation in the quantum theory.*
>
> Take as an example a Wilson cloud chamber ... experiment. The causal chain here consists of the formation of drops of water round ions, the scattering of light by these drops of water, and the action of this light on a photographic plate, where it leaves a permanent record.... One could perhaps extend this chain further into the past. In general one tries with the help of theoretical considerations to extend the chain as far back into the past as possible, in order that the numbers obtained as the result of the experiment may apply as directly as possible to the process under investigation.

This view of the nature of the results of experiments fits in very well with the new quantum mechanics.... the state of the world at any time is describable by a wave function ... , which normally varies according to a causal law, so that its initial value determines its value at any later time.

Now, he goes on to describe an expansion of the wave function into a basis set derived from the possible outcomes of the experiment, with a probability given by the squared magnitude of the coefficients of this expansion. The particular selection of any one basis function is an "irrevocable choice of nature, which must affect the whole of the future." He then discusses a scattering event in which placing a reflecting mirror in the system may upset the chain of events and present the problem of not knowing which possible path (reflected or not reflected) nature has chosen. That is, the interference of the various basis functions, due to the mirror "compels nature to postpone her choice." The comments of Bohr following Dirac's comments are not sufficiently well collected to tell us what his point was, but he does remark that the experiment plus all the apparatus including the photographic plate still makes a closed system. Kramers points out that the natural approach with the wave function leads to the probability that Dirac discussed, but Heisenberg objects to Dirac's point that "nature makes a choice." His argument is that if one is far away, the fact that nature had made a choice makes it hard to understand how the interference was produced. He would prefer that the observer had made the choice. Lorentz summarized the discussion by stating that there appeared to be a fundamental difference of opinion on the choices that could be made by nature.

There then followed a discussion on radiation pressure produced by photons striking a surface. Kramers resurrected the issue from a comment Brillouin made during de Broglie's lecture. There followed discussion by de Broglie, Kramers, and Brillouin. At one point, Kramers asks what advantage can one gain by having a precise velocity for the photon. De Broglie pointed out that this allowed one to consider trajectories for the photons that had a positive meaning, where Kramers replied that he saw no advantage in experiments by having a picture that include well-defined photon trajectories.

Einstein then shifted the argument to a discussion of the photon velocity during its travel in the interference zone, which does not seem to have been resolved, other than to say that the photon traveled at the speed of light outside this zone.

Langevin raised the question of which statistics to use. He suggested that perhaps molecules would satisfy Bose–Einstein while particles would require Fermi–Dirac. This was suggested to imply that material particles were distinguished from photons by their impenetrability. Heisenberg corrected him to point out that, in quantum mechanics, there was no reason to choose one form of statistics over another. He did remark that Bose–Einstein were better for photons, while positive and negative electrons were better described by Fermi–Dirac. Kramers pointed out that Dirac had shown that Bose–Einstein statistics could be derived from the black-body cavity, but Dirac replied that this separation is quite artificial. In fact, Dirac noted that the manner in which the two set of statistics are derived is quite different. Kramers then noted that no one seems to understand this, and especially doesn't understand why Dirac felt the need to quantize the Schrödinger waves. No one it seems felt compelled to point to the new understanding of spin as the leading difference between the two forms of statistics.

Then, the discussion diverged into that of black-body radiation and the absorption and emission bands of helium. Ehrenfest brought them back to the fundamental questions by posing an experiment dealing with the Pauli exclusion principle. Suppose that we have a box with a small opening so that non-interacting particles (before entering the box) are allowed to enter the box, and stay there for a long period before leaving. If they are monoenergetic upon entering, will they have a spread of energy upon exiting due to the interactions among the particles in the box. Heisenberg replied that two electrons must have different energies, but if the interaction is very small, it takes a long time to establish the required exchange of energy between themselves. Then, Richardson pointed out that one also needed to consider nuclear spin, for which the evidence is much more complete than Heisenberg had considered. Langmuir again brought them back to the main issue by asking how similar the analogy between photons and light waves is to the analogy of particles and de Broglie waves. Ehrenfest then pointed out the question of how to describe linearly polarized optical waves with photons. He pointed out that for electrons, spin may be described by an antisymmetric tensor which the particles carry with their motion. Compton then asked whether light could be elliptically polarized if the photon carried angular momentum, but Ehrenfest replied that he didn't really understand what it means to assign a universal angular momentum to a photon.

Pauli then pointed out [1]:

> The fact that the spinning electron can take two orientations in the field allowed by the quanta seems to invite us at first to compare it to the fact that there are, for a given direction of propagation of the light quanta, two characteristic vibrations of blackbody radiation, distinguished by their polarization. . . . These two modes of description are mathematically equivalent, but independently of whether one decides in favor of one or the other, it seems to me that one cannot speak of a *simple* analogy between the polarisation [*sic*] of light waves and the polarization [*sic*] of de Broglie waves associated with the spinning electron.

Dirac then pointed out a failure in the analogy between the spin of electrons and the polarization of photons. The spin variable of an electron commutes with the dynamic variables of momentum and position, but the case is different for photons. While one could give a polarization for a plane wave in a manner that the polarization would commute with the momentum, if the position of a photon is localized one cannot give a definite polarization to it.

Lorentz then commented that the classical electromagnetic waves allowed one to create an energy–momentum tensor, which survived in relativity. But if you used photons, there would be a question about this ability. That is, if one assumed that the photons all traveled at the speed of light, it appeared that photons could not be used to give this same tensor. De Broglie replied that he had not worried about this when he related the motion of photons with his new quantum mechanics, as he considered only waves of scalar character. Lorentz then questioned whether his waves were different from light waves, to which de Broglie replied that one does not know yet all the physical character of his waves. Discussion about the speed of the photons in, and away from, the interaction zone led de Broglie to point out that, in the corpuscular theory of light, diffraction led to the necessity for curved trajectories for the photons. This leads to the need for new considerations for the forces in the interaction zone.

Pauli then turned to a problem with the de Broglie approach. He first pointed out that the ideas of de Broglie were in full agreement with Born's approach in the case of elastic collisions, then asserted that this was not the case for inelastic collisions. He used the case of a "rotator," or oscillator moving in a plane, which had been discussed by de Broglie himself. Pauli brought out a treatment by Fermi which demonstrated that a collision of a particle moving in the same plane as the rotator and the latter. If there is no interaction, then Born had shown that one needed to use the superposition of several partial waves to account for the possible states of the rotator. Yet, de Broglie had claimed that the rotator no longer possessed a constant angular velocity, which brought the two approaches into disagreement. If now one introduced an interaction between the particle and the rotator, Fermi had shown that the energy of interaction depended upon the angle of the rotator, and this had a nice analog in a wave falling on a grating (because of the periodicity of the angle of rotation). Here, Pauli pointed out the essential point that if the rotator was in a stationary state before the interaction, the incident wave must be of infinite extent in the direction of the axis; e.g., a proper plane wave. For this reason, the different spectral orders of the grating will all be superimposed at each point in configuration space (of the two coordinates and the angle of rotation). Then, the crucial point becomes [1]:

> If we then calculate, according to the precepts of Mr. de Broglie, the angular velocity of the rotator after the collision, we find that this velocity is not constant. If one had assumed that the incident wave is limited in the direction of the axis [of the angle], it would have been the same before the collision. Mr. de Broglie's point of view does not then seem to me compatible with the requirement of the postulate of quantum theory, that the rotator is in a stationary state both before and after the collision.

He then went on to say that this incompatibility was not unique to the rotator example, and was due to the condition assumed by de Broglie that the behavior of the particles would be completely determined while also satisfying ordering kinematics at the same time. That is, the assumption of a close connection to classical mechanics by de Broglie was incompatible with the results one expected from quantum mechanics. This was a dagger directly to the heart of de Broglie's approach. De Broglie responded by pointing out that Fermi's problem was not really of the same type as Pauli's example and that similar difficulties had their analogs in the classical optics of diffraction. This elicited further discussion, but the damage had been inflicted, and it did not seem that de Broglie was providing the necessary damage control.

At this point, Lorentz took the discussion in another direction, perhaps recognizing that the damage inflicted would lead to more arguments than discussion. He returned to the hydrogen atom, and asked about the understanding for a single eigenstate of the electron with a given energy. He then asked whether squared magnitude of the eigenfunction for that state expressed the probability of that level. The issue he remarked was that the electron could not escape from that eigenstate where as the squared magnitude of the wave function had value at very large distances. This he felt was "disagreeable." De Broglie responded that in the relativistic form, the electron was no longer constrained to remain within the region of the atom, which was perhaps more disagreeable. Born then felt compelled to bring Schrödinger into the argument as well, and remarked that, in opposition to the ideas of Schrödinger, it was nonsense to talk about positions and movement of an electron in the atom, in the absence of experiments that could measure them. Pauli thought it was no problem to determine the location of the particle outside the atom even without sufficient energy to cause the ionization of that state.

Lorentz then threw in another question. If one took the wave mechanics view, the wave function for the electron was smeared out over the entire atom, which in a manner of speaking was also the view in matrix mechanics. If one tried to create a localized wave packet for the electron, this seemed to be difficult in the atom. Moreover, the wave packet does not retain its shape, and spreads with time. This picture of the electron being described by a wave packet was not satisfying to him except on the short time scale. He then noted that Bohr's approach would re-localize the packet after an observation, and then the time scale over which the packet would spread would restart. That was nice, but for his part, Lorentz felt he needed a description which covered the entirety of time in a smooth manner. Schrödinger responded that he had no problem with the lack of a packet, and that this lack of the packet was precisely the important point in wave mechanics, and was the absolute failure of the old classical view of mechanics. Born added that determinism also required that the future location of a particle required great accuracy of the knowledge of the initial point of the particle, and classical mechanics failed because the laws of the propagation of any such packets were different. Thus ended the discussion sessions.

Solvay 1927: Einstein and Bohr

The above discussion of the actual conference discussion is long, and often tedious. But it is important as well for what it does not contain. And this is the sideline arguments between Einstein and Bohr. These were alluded to above, and discussed enthusiastically by Heisenberg as mentioned. At this conference, they primarily concerned uncertainty and Heisenberg's uncertainty principle. One suspects that there were at least three such problems, and we know that Bohr responded to two of these. Bohr did not respond to the third one, on the α particles, as Pauli discussed this problem during the formal discussion period as noted above. If there were more, they seem to have been lost in the intervening years. So, let us discuss these two problems. They both involve passing particles through a slit, and so are related to the problem that Einstein posed during the general discussion. We should perhaps consider them as preliminary to that discussion.

In the first problem, Einstein considered that we would send a beam of particles through a small hole in a screen, S_1, as shown in Fig. 4.1 [35]. This hole would be sufficiently small that the particle, or its wave, would be diffracted by the act of passing through the hole, and the particle would then gain a momentum in the up

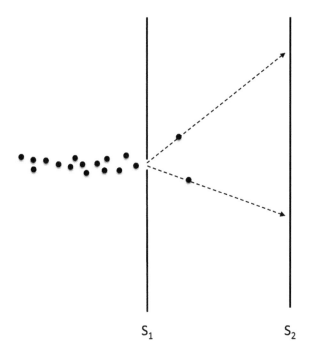

Figure 4.1   Einstein's first problem for Bohr, which is discussed in the text.

or down direction. This would cause it then to reach the second screen at a point not in line with the hole. In addition, though, he placed a shutter on this hole, so that the particle could only pass through for a short period of time. Hence, there would be a limited number of particles between the two screens. Now, the idea is to study the points of arrival of the particles on a second screen, S_2, perhaps by placing a photographic plate at this second screen. From the point of arrival, Einstein argued that he would now know the exact momentum given to the particle in the up or down direction and would also partially know the position of the electron. Thus, this should violate Heisenberg's uncertainty relation. But Bohr saw through this relatively quickly. He pointed out that Einstein assumed that the screen remained stationary and that there was no uncertainty in the shutter. The screen S_1 and the shutter are actually coordinated by the space-time reference frame in Einstein's own theory of relativity. Moreover, the screen S_1 would experience a recoil momentum when the particle was diffracted, hence there was an uncertainty in its position as well as in the amount of momentum imparted to the diffracted particle. Einstein's conclusion could only be supported if both screens were of infinite mass. Thus, there was an uncertainty inherent in the experiment, and this could not be avoided.

So Einstein modified the configuration. He considered that two slits were placed in the second screen and that a new third screen would be used for the measurement. The second screen S_2 would be mounted on a spring so that the recoil momentum given to this screen by the diffracted particle would depend upon which slit it passed through. By reducing the particle density to the point at which only a single particle at a time was in the system, one still would be able to see the wave properties by the resultant interference pattern that would build up. But one could also determine just which slit the particle passed through by monitoring the motion of the second screen. Bohr, however, deflected this new experiment as well. Since, from the previous example, there was an uncertainty in the amount of the momentum transferred to the particle, and the resulting recoil momentum of the second screen, there would be an uncertainty in the position of the two slits. This would defeat the accuracy of the experiment. Thus, this could not defeat the uncertainty principle.

To the casual observer, such as Paul Ehrenfest, it appeared that Einstein had been defeated, and that he should admit that he was wrong. But there is an alternative interpretation that usually is not considered. Einstein had sat with Bohr for a week in Leiden. It is unreasonable to assume that he had not come to the same conclusion as John Slater. Contrary to the views of some, this conclusion was that Bohr was not a deep thinker, and preferred to obfuscate the major points of a discussion and announce his de facto result at the end. Even his brother Harald had said [36],

> When Niels tells me something, I absolutely don't understand what he is talking [about] and what he is driving at for 59 minutes; but in the 60th minute a light suddenly dawns and I see that everything he had said previously was absolutely necessary.

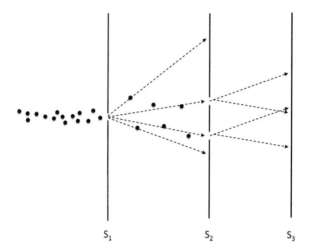

Figure 4.2 Einstein's modified setup for his second problem, discussed in the text.

Perhaps Harald was correct, but it may as well be the case that in the last minute Bohr told him what the outcome should be and Harald just assumed that the rest of the discussion supported that. Ehrenfest was a little closer to the truth [36]:

> Ask and then listen patiently and attentively to what follows, and just don't come back to the question. In this way I learned an enormous amount, but I had hardly a tenth of my questions "answered."

Einstein would have concluded from this that it might just be easy enough to draw Bohr into a trap. One may notice immediately that Bohr's responses to the two problems above say absolutely nothing about quantum mechanics. Bohr had taken great pride in using Einstein's relativity theory to prove him wrong, but at the same time, he used nothing but the classical problems of making simultaneous position and velocity measurements. We discussed earlier the problem of measuring train cars as one waited patiently at a blocked railroad crossing. Uncertainty is not limited to quantum mechanics. In fact, in most circumstances, the Heisenberg uncertainty is not more than a factor of 2 smaller than the classical uncertainty. This can be shown simply by considering an energy time problem, which is purely classical. Consider a voltage which decays exponentially as $\exp(-t/\tau)$. If we use this as the probability function, then the expectation value of t is τ. In other words, in order to measure the nature of the time variation, we have to measure for at least τ seconds. What bandwidth does this require on the measuring apparatus? We answer this by the one-sided Fourier, or the Laplace, transform. In frequency space, the spectrum is flat up to an angular frequency of $1/\tau$. Hence, our bandwidth has to be at least this wide. As a result, we can write $\Delta\omega\Delta t = (1/\tau) \cdot \tau = 1$. If we now use Planck's relation, we get $\Delta E\,\Delta t = \hbar$. Thus, this classical problem has an uncertainty that is only a factor of 2 from Heisenberg's quantum mechanics' minimum uncertainty. It is only this factor of 2 that arises from the quantum approach with noncommuting operators.

Thus, Bohr had failed to show that quantum uncertainty was necessary to explain the problems. He had fallen into Einstein's trap, and had not dispelled Einstein's claim that quantum uncertainty was an avoidable limitation on knowledge. Only Pauli, in his response to the α-particle problem, during the general discussion, did so with quantum mechanics, and this was through wave function collapse rather than resorting to anything related to the uncertainty principle. The idea that Bohr had won the debate by showing a need for uncertainty missed the point. He had not defended the quantum uncertainty at all in his response.

The Aftermath

Barely two months after the conference, Hendrik Lorentz became seriously ill and died a few days later. He was 74. He had presided at all of the Solvay conferences, but he was not an idle spectator. It was remarked earlier that he had encountered Schrödinger during a trip to America and that perhaps this had helped with Schrödinger's invitation to the conference. In fact, Lorentz gave a series of lectures on Schrödinger's wave mechanics at Cornell during the fall of the preceding year, so he was familiar with the developments. He was also familiar with Schrödinger, as the latter had attended the 1924 Solvay conference, and the two had significant correspondence during 1926. Lorentz's opening remarks in the discussion session were very insightful, and it was clear that he had serious doubts about the Bohr and Heisenberg philosophy of the experiment defining reality. It was also clear that he considered the present state of quantum mechanics to represent an incomplete theory. In these views, he sided with Einstein. He was a respected scientist and individual. Richardson [37] would write of him:

> He possessed and successfully employed the mental vivacity which is necessary to follow the interplay of discussion, the insight which is required to extract those statements which illuminate the real difficulties, and the wisdom to lead the discussion among fruitful channels, and he did this so skillfully that the process was hardly perceptible.

Klein [38] would say:

> For many years physicists had always been eager "to hear what Lorentz will say about it" when a new theory was advanced, and, even at seventy-two, he did not disappoint them. His letter of May 27, 1926 [to Schrödinger] consists of a long analysis and critique of Schrödinger's work in which Lorentz puts his finger on some of the most questionable points.

At the end of the conference, Lorentz had to spend time working on his Como manuscript, which he had put off writing. As a result, he left the task of editing the transcript of the Solvay conference to the secretary, Verschaffelt. He died before he could read the result.

Most of the editing of the final version of the proceedings was done solely by Verschaffelt. The death of Lorentz allowed Bohr to perpetrate one of the worst scientific hornswoggles of the age. Bohr induced Verschaffelt to remove most of his contributions to the general discussion and to replace it with his Como lecture, as it appeared in the German form [39]. It appeared with the following footnote [1]:

> This article, which is the translation of a note published very recently in Naturwissenschaften, vol. 16, 1928, p. 245, has been added at the author's request to replace the exposition of his ideas

that he gave in the course of the following general discussion. It
is essentially the reproduction of a talk on the current state of
quantum theory that was given in Como on 16 September 1927,
on the occasion of the jubilee festivities in honour of Volta.

While it is thus clearly marked as *not being* one of the lectures,
this footnote and the paper have been greatly misunderstood by
following generations.

Placing the paper on the same footing as the lectures leads
subsequent readers to assume that the Committee thought that Bohr
should be a lecturer along with Bragg, Compton, de Broglie, Born
and Heisenberg, and Schrödinger. They did not! The validity of this
comment lies in the fact that at least one modern writer on quantum
mechanics did come to this conclusion [35], as he refers to Bohr's
talk.

It leads one to believe that the Committee felt that complementarity was a great scientific advance on the same level as
matrix mechanics, de Broglie's wave and particle approach, and
Schrödinger's wave mechanics. They did not.

It leads one to believe that there was a general agreement that
complementarity was an accepted scientific theory by most of the
attendees. This was not the case even among Bohr's close fraternity.
Even in the Como talk, Bohr knew that Heisenberg was not favorably
disposed to the ideas [40]:

> It is unlikely that Heisenberg missed the barbs directed at him
> by Bohr—or the inclusion of concepts Bohr knew Heisenberg
> opposed—even if everyone else in the audience did.

Heisenberg [27] himself said, much later when he had softened on
the idea, that

> it was already quite clear in those years that the two pictures, wave
> and particle picture, are the two most important pictures which we
> have.... I don't object nowadays very much against the dualistic
> description if only people keep in mind that actually they mean
> only one theory.

It certainly wasn't accepted by others, such as Einstein and Lorentz.
In his paper, Bohr had, of course, declared that the quantum world
was not visualizable, as we have discussed. This meant to them that

Bohr has just gotten rid of any need to try to explain any remaining paradoxes. As Jammer [41] puts it, the wave–particle idea,

> which he [Bohr] subsequently called the quantum postulate according to which an essential discontinuity or not-further-analyzable individuality, a notion completely foreign to classical thought and symbolized by Planck's quantum of action, has to be attributed to every atomic process.

Following Bohr's remarks in the general discussion (since redacted to be replaced by his paper), Einstein gave the theory a serious rejection, which set off a disordered general response best described by the actions of Ehrenfest, who wrote upon the board the passage from Genesis [1]:

> And they said one to another: Go to, let us build us a tower, whose top may reach unto heaven; and let us make as a name. And the Lord said: Go to, let us go down, and there confound their language, that they may not understand one another's speech.

Certainly, Bohr's introduction of complementarity and his philosophy into the proceedings of the conference created such a result for quantum mechanics. Yet, some of his disciples would move on to legitimize the entire idea. Pauli, in a later review would suggest that the notion of complementarity was the defining characteristic of quantum mechanics [42]. Thus, the Copenhagen Interpretation became the official dogma for the next third of a century. It is perhaps more troubling that it remains close to the heart in many, if not most, physicists today.

The year following the conference would be difficult for many. Einstein fell ill due to heart problems and suffered a long convalescence. Born would undergo a physical, possibly mental, breakdown and undergo an even longer convalescence. Pauli would move to Zürich and Jordan would take his position in Hamburg. Schrödinger would spend an inordinate amount of time on his other preoccupation—L'amour. Perhaps the worst of this is that de Broglie, perhaps disillusioned with his own theory, began to teach and use matrix mechanics. He would not return to his own theories until after the coming conflagration, when David Bohm would accept the concept of complementarity but not the philosophy that came with it. Yet, Holton [43] has said, "A distinguished group of physicists, although a minority in the field, remained unconvinced by and indeed hostile to the complementarity point of view." Even Born would write to Einstein [44]:

> You are, of course, absolutely right that an assertion about the possible future acceptance or rejection of determinism cannot be logically justified. There can always be an interpretation which lies one layer deeper than the one we know.

Einstein would write to Schrödinger [45] many years later:

> You are the only contemporary physicist, besides Laue, who sees that one cannot get around the assumption of reality—if only one is honest.

References

1. G. Bacciagaluppi and A. Valentini, *Quantum Theory at the Crossroads: Reconsidering the 1927 Solvay Conference* (Cambridge Univ. Press, Cambridge, 2009).

2. P. P. Ewald, *Phys. Today* **29**(7), 47 (1976); this is a review of J. Mehra, *The Solvay Conferences on Physics: Aspects of the Development of Physics since 1911* (Riedel, Dordrecht, 1975).

3. P. Langevin and M. de Broglie, *La Théorie du Rayonnement et les Quanta* (Gauthier-Villars, Paris, 1912).

4. N. Straumann, *Eur. Phys. J.* **36**, 379 (2011).

5. P. Debye, *Ann. Phys.* **33**, 1427 (1910).

6. G. Holton, *Thematic Origins of Scientific Thought: Kepler to Einstein*, 2nd (revised) edition (Harvard Univ. Press, Cambridge, 1988), p. 207.

7. S. Jones, *The Quantum Ten* (Thomas Allen, Toronto, 2008), pp. 22–23.

8. A. Einstein, in *The Collected Papers of Albert Einstein*, Vol. 5, correspondence 1902–1914, ed. by M. J. Klein, A. J. Knox, and R. Schulmann (Princeton Univ. Press, Princeton, 1993).

9. S. Jones, *op. cit.*, p. 66.

10. P. Debye, in *Lectures on the Kinetic Theory of Materials* (Teubner, Leipzig, 1914), p. 27.

11. A. H. Compton, *Phys. Rev.* **21**, 483 (1923).

12. P. Debye, *Phys. Z.* **24**, 161 (1923).

13. P. Debye, in American Institute of Physics Oral Histories: https://www.aip.org/history-programs/niels-bohr-library/oral-histories/4568-2

14. L. de Broglie, *J. Phys. Rad.* **8**, 225 (1927); trans. by W. M. Deans in L. de Broglie and L. Brillouin, *Selected Papers on Wave Mechanics* (Blackie & Sons, London, 1928), pp. 113–138.

15. J. T. Cushing, *Quantum Mechanics, Historical Contingency and the Copenhagen Hegemony* (Univ. Chicago Press, Chicago, 1994), pp. 126–128.

16. S. Jones, *op. cit.*, p. 238.

17. J. S. Bell, *Found. Phys.* **12**, 989 (1982); reprinted in J. S. Bell, *Speakable and Unspeakable in Quantum Mechanics* (Cambridge Univ. Press, Cambridge, 1993), pp. 159–168.

18. A. Einstein, *Bestimmt Schrödingers Wellenmechanik die Bewegung eines Systems vollständig oder nur im Sinne der Statistik?*, unpublished.

19. D. W. Belousek, *Stud. Hist. Phil. Mod. Phys.* **27**, 437 (1996).

20. T. Kaluza, *Sitz. Pruss. Acad. Wissen*sch., 22 December 1921, pp. 966–974.

21. A. Einstein, *Sitz. Pruss. Acad. Wissen*sch., January 1927, pp. 23–25.

22. P. Ehrenfest and G. E. Unlenbach, *Proc. Akad. Wetensch. Amsterdam* **29**, 1280 (1926).

23. O. Klein, *Z. Phys.* **37**, 895 (1926).

24. S. Jones, *op. cit.*, p. 289.

25. P. McEvoy, *Niels Bohr: Reflections on Subject and Object* (Microanalytix, San Francisco, 2000), p. 113.

26. L. Rosenfeld, in *Physics in the Sixties*, ed. by S. K. Runcorn (Oliver and Boyd, Edinburgh, 1963), pp. 1–22.

27. W. Heisenberg, in American Institute of Physics Oral Histories: https://www.aip.org/history-programs/niels-bohr-library/oral-histories/4661-9 and also 4661-11.

28. A. Pais, *Niels Bohr's Times, in Physics, Philosophy, and Polity* (Clarendon Press, Oxford, 1991), p. 433.

29. P. A. M. Dirac, *Proc. R. Soc. A* **114**, 293 (1927).

30. P. A. M. Dirac, *Proc. R. Soc. A* **114**, 710 (1927).

31. P. McEvoy, *op. cit.*, p. 79.

32. G. N. Lewis, *Nature* **118**, 874 (1926).

33. G. I. Taylor, *Proc. Camb. Phil. Soc.* **15**, 114 (1909).

34. N. Bohr, *Niels Bohr: Collected Works*, Vol. 6, ed. by J. Kalckar (North-Holland, Amsterdam, 1985).

35. A. Whittaker, *Einstein, Bohr, and the Quantum Dilemma* (Cambridge Univ. Press, Cambridge, 1996), pp. 202–210.

36. M. J. Klein, *Physica* **106A**, 3 (1981).

37. O. W. Richardson, *J. London Math. Soc.* **4**, 183 (1929).

38. M. J. Klein, in *Letters on Wave Mechanics: Schrödinger, Planck, Einstein, Lorentz*, ed. by K. Przibram (Philosophical Library, New York, 1967), pp. ix–xv.

39. N. Bohr, *Naturwissen.* **16**, 245 (1928).

40. S. Jones, *op. cit.*, p. 246.

41. M. Jammer, *The Conceptual Development of Quantum Mechanics* (McGraw-Hill, New York, 1966), p. 347.

42. W. Pauli, *Handbuch der Physik*, Vol. 24, ed. by H. Geiger and K. Scheel (Springer, Berlin, 1933), p. 126.

43. G. Holton, *op. cit.*, p. 104.

44. M. Born, letter to A. Einstein of January 13, 1929, reprinted in *The Born-Einstein Letters*, trans. by I. Born (Macmillan, London, 1971).

45. A. Einstein, letter to E. Schrödinger of December 22, 1950, reprinted in *Letters on Wave Mechanics: Schrödinger, Planck, Einstein, Lorentz*, ed. by K. Przibram (Philosophical Library, New York, 1967), pp. 39–40.

Chapter 5

Interregnum

The period following the 1927 Solvay conference was one of comparative peacefulness. This does not mean that there were no advances; there certainly were many important advances, and the first would be led by Paul Dirac. It would be a few years before the next, highly discussed structural challenge would be made to the Copenhagen orthodoxy. And, once more, this would come from Einstein. Einstein and Bohr would continue to spar during this period, but this would just provide the build up to the big challenge of Einstein–Podolsky–Rosen in 1935. In this chapter, we want to look at this sparring and concentrate on the very important developments in quantum theory that occurred during the interregnum. We begin with Paul Dirac.

Dirac Moves Forward

During the just concluded Solvay conference, Dirac had made very few comments, as was his nature. However, he had made one following Compton's talk that was significant if the attendees had been paying close attention. Here, he had commented that "the rate of change of the momentum of the electron ... is equal to $1/c$

The Copenhagen Conspiracy
David Ferry
Copyright © 2019 Pan Stanford Publishing Pte. Ltd.
ISBN 978-981-4774-75-8 (Hardcover), 978-1-351-20723-2 (eBook)
www.panstanford.com

times the ... rate of change of the energy due to the radiation" [1]. To the cognoscenti, this remark should have created some interest and raised a few eyebrows. Here, Dirac almost gave his game away. So, in the general discussion, he further remarked, "At present the general theory of the wave function ... involves the abandonment of relativity" [1]. This was a smoke screen. In further support of this smoke screen, he mainly commented on the matrix theory, and its comparisons to a classical approach in which one considered the dynamical variables to be operators. He already had produced his own development of matrix mechanics, which appeared at the time of the Born-Jordan and Drei Männer papers. At the time of this Solvay conference, he was working hard on connecting quantum mechanics to relativity theory. In this pursuit, he had already concluded that matrix mechanics was not up to the job, and that he could make progress with wave mechanics. He had, in fact, succeeded in just the task he claimed could not be done, and the comment after Compton's talk was an indication of that. But he still needed a few weeks to complete the manuscript, and he did not want these guys playing in his sandbox. He did not want to signal the results before the paper appeared in print. Following the conference, the paper was submitted near the end of December by Fowler.[a] It was received by the journal on January 2 and appeared in the February 1 issue [2]. It would knock Pauli off his chair. So far, any consideration of spin was an add-on, whether it was matrix mechanics or wave mechanics. Now, spin was an integral, and seminal, part of the theory, and it would be wave mechanics that made it possible.

The need to develop a relativistic form of quantum mechanics was realized almost from the beginning, at least by Schrödinger, who proposed an extension of his famous equation that meets the requirements of special relativity, but in this form he had to use the second derivative with respect to time in the equation [3]. He had actually developed the equation, which met the requirements of relativity theory, but didn't work on the atomic problem. So, he dropped it and started the efforts leading to his better known

---

[a]The premier journal in Britain was the *Proceedings of the Royal Society*. Papers here needed to be submitted by one of the Fellows, which Fowler was.

equation. He finally published this earlier work after the series of four papers on wave mechanics. Others worked on the same ideas in 1926, and the equation became known as the Klein–Gordon equation [4, 5]. The equation worked well for photons, because it was easy to incorporate the electromagnetic potentials coming from Maxwell's equations, but there was no way to incorporate spin into the solutions. It survives today for studies of relativistic effects with photons, but these authors began at the same place as Dirac, and that is with Einstein's relativistic relationship between energy and momentum

$$E^2 = p^2 c^2 + m_0^2 c^4 . \tag{5.1}$$

Now, the significance of Dirac's comment following the Compton talk is clear. To mention the factor of $1/c$ pointed directly to relativity theory. Sure, there were many people interested in the problem of connecting this to quantum mechanics, but the spin was the central issue. This equation has interesting implications. First, there are both positive energy and negative energy solutions when one solves for just the energy E. These two solutions are symmetric about a gap (where the momentum $p = 0$) of $m_0 c^2$. Here, m_0 is the rest mass of the particle, and c is the velocity of light. No one quite knew what the negative energy solution meant, so they worked only with the positive energy solution that they understood.

Dirac recognized that the first term on the right-hand side represented a kinetic energy, and saw that an equation for the energy itself must be linear in the momentum. This was an enormous leap from classical mechanics, but this was what Einstein's relativistic equation (5.1) demanded. Dirac approached this by formulating the simplest Hamiltonian that is linear in the momentum and mass terms (it was a master stroke to realize this equation would work), which is

$$H = c\tilde{\alpha} \cdot \mathbf{p} + \beta m_0 c^2 . \tag{5.2}$$

Here, $\tilde{\alpha}$ is a three-dimensional vector and β is some constant. But now if we make the same connections to time and space derivatives, since $\mathbf{p}$ is an operator, a wave equation results as

$$i\hbar \frac{\partial \psi}{\partial t} = -i\hbar c\tilde{\alpha} \cdot \nabla \psi + \beta m_0 c^2 \psi . \tag{5.3}$$

This is a wave equation of a new type, but one which is fully relativistic. It became known as the Dirac equation, and creates relativistic wave mechanics. In the first paragraph, he points out this new quantum theory of the electron was fully compatible with Pauli's spin [6], but what were these new parameters? As mentioned, $\bar{\alpha}$ is a three-dimensional vector, but the symmetry between positive and negative energies required that this quantity, plus the scalar, must satisfy certain commutator products, just as noncommuting operators were required in both matrix mechanics and Schrödinger's wave mechanics. That meant four quantities—the three components of $\bar{\alpha}$ and β.

Now, in Pauli's world, the normal wave function was modified with a multiplicative spin function, and this spin function described the up- or down-spin with a two component matrix, as

$$\varphi(\uparrow) = \begin{bmatrix} 1 \\ 0 \end{bmatrix}, \quad \varphi(\downarrow) = \begin{bmatrix} 0 \\ 1 \end{bmatrix}, \tag{5.4}$$

so that the direction of the arrows told us in which direction the spin pointed. He then developed a set of 2×2 matrices that transformed one spin into the other. There were 3 such transformation matrices, conveniently called σ_x, σ_y, σ_z. So, in a sense these would work for his new vector parameter. However, there were only 3 of these matrices. With β, Dirac needed 4 such matrices. With matrices of rank 2, there are only the three possibilities given by the Pauli matrices, if you demand that they do not commute. To get four such matrices, you have to go to rank 4 matrices, but Einstein had already claimed that relativity theory put space and time on equal footing. This gives 4 dimensions, so that using rank 4 matrices for these new constants was consistent with relativity theory. Moreover, he could show that these four new matrices were block matrices composed of 2×2 blocks, and these smaller blocks were the Pauli matrices in various forms. So, if the Pauli matrices fit naturally into the theory, then spin was an integral part of the theory and the Dirac equation. On the other hand, if the matrices were rank 4, then there were 4 solutions to the Dirac equation. He needed only two solutions for the spin up and spin down electron. What were the other two solutions?

We remarked above that Einstein's energy equation had positive and negative energy solutions. Dirac's two solutions for the spin up and spin down electrons were obviously the positive energy solutions. There were then two other solutions, which corresponded to negative energy, and *possessed a spin in the same manner as the electron solutions*. But Dirac generally, and casually, assumed the negative energy states were all full and didn't affect the positive energy results. Then, he applied his result to the atomic problem where the electron moved in a central, spherically symmetric, potential and showed how spin and normal angular momentum could be handled, as well as showing how this approach also yielded the Zeeman effect [7]. But what did these negative energy solutions mean?

Following this work, Dirac turned his attention to the many-body problem that naturally would arise in multi-electron atoms. He based his approach upon one that Hartree had discussed earlier [8] (this work had been mentioned already at the Solvay conference the previous year). Although Dirac worked out the mathematics of permuting the order or position of electrons, using Hartree's approach was only a mean field approach to the problem. Thus, he missed the asymmetry that would be introduced with permutations in later years via what is called the exchange interaction, directly related to these permutations, but he would work on this interaction between electrons for some time. Heisenberg and Pauli published an approach to understand how the electrons interacted with the electromagnetic field [9], where they regarded the electromagnetic field as a dynamical variable amenable to use within a Hamiltonian approach. Dirac found this objectionable: "We cannot therefore suppose the field to be a dynamical system on the same footing as the particles and thus something to be observed in the same way as the particles. The field should appear as something more elementary and fundamental" [10]. He then went on to do the quantization of the electron and the field sequentially, and reached the conclusion that "the interactions between particles takes place by means of vibrations of an intervening medium transmitted with a finite velocity." Today, we recognize this as the emission of a virtual photon by one electron and its subsequent absorption by the second electron. The natural carrier of the Coulomb potential between the two electrons is part of the electromagnetic field, and the carrier of forces due to this potential is seen as the photon.

The question of the understanding of the negative energy states was never far from the surface at the time. Dirac [11] pointed out that

> an electron with negative energy moves in an external potential as though it carries a positive charge. This has led people to suspect a connection between the negative-energy electron and the proton or hydrogen nucleus [12].

Dirac, however, felt that the interaction between an electron at positive energy and the proton at negative energy would not conserve energy and momentum. He then subsequently suggested [13] that

> we can assume that in the world as we know it, *all* ... of the negative-energy states for electrons are occupied. A hole, if there were one, would be a new kind of particle, unknown to experimental physics, having the same mass and opposite charge to an electron. We may call such a particle an anti-electron.

Today, this anti-electron is called the *positive electron*, or positron. Dirac had just opened the scientific community to the likely existence of antiparticles and pointed out that they were required by the four solutions to his relativistic quantum mechanics. It was interesting that this comes about three years after the Dirac equation. Dirac [14] would say later:

> I felt that writing this paper on the electron was not so difficult as writing the paper on the physical interpretation. That really needed more concentrated and sustained thought to get the ideas clear.... The negative energies were a problem right from the beginning but one just couldn't do anything about them.

Heisenberg [15] later remarked,

> I do remember that I felt very uneasy about it. I saw, "Well, this Mr. Dirac has not only introduced one doubling—that of the spin, which we know and which Pauli had explained to us was necessary—but he had to introduce another doubling, that

between positive and negative energies." I felt at once that now we are getting into trouble. I think not only I, but also many other people would feel that something has happened which is disagreeable. It was the first time that there could creep in the feeling that even with quantum mechanics not all problems are solved.

And Bohr? According to Pais [16],

> It took Bohr a rather long time to accept Dirac's ideas. In a lecture given in October 1931, after the positron proposal [but before its experimental finding], he referred to "the peculiar difficulties which present themselves in the attempt to develop a proper relativistic treatment of atomic problems."

And again from Heisenberg [15]:

> Bohr was initially very skeptical about the existence of a particle at all but that he also insisted finally that maybe it exists but it's got nothing to do with this stuff of Dirac's, with Dirac's hole.

Dirac's development of a formulation for coupling quantum theory and relativity was remarkable, and a singular achievement. It may have been as ground breaking as the original work by Heisenberg, de Broglie, Pauli, and Schrödinger. The work opened the door to antimatter, and then subsequently to particles more fundamental than even the electron. The positron is thought to have been observed first by Skobeltzen [17], but Anderson is credited with the first convincing evidence for its existence [18]. Perhaps even more important is the conclusion one must draw from this advance, and the remaining unanswered questions, that it is simply impossible to conclude that quantum mechanics is complete, even today. *Moreover, Dirac's developments are the result of his "beautiful" mathematics, and the results are correct and predictable, even without the experimental observations. It was a challenge to Bohr's ideas, since, if Dirac were right, reality existed even before any measurement, especially as measurement came long after the theory.*

But despite Bohr's objections, relativistic quantum mechanics continued to move forward. Shortly after Dirac's work, Weyl suggested that one could get Dirac fermions with one-half the charge of the electron [12]. These particles have been studied in some models of quantum field theory, and there is some indication in very recent years that they may have been seen in some optical experiments. One notable follow-on to Dirac's papers arrived in 1937 from Ettore Majorana. Majorana showed that one could rearrange Dirac's equation into a form that did not need to have the full negative-energy states, but more importantly, there was a solution at zero energy which supported a particle that was its own antiparticle [19]. Today, we call this a Majorana fermion, and there has been a steady search for the particle. More attention has been given to the man, as he apparently withdrew all his cash from the bank in 1938 and boarded a ship going from Palermo (in Sicily) to Naples, only to never be seen again. He did leave a note, but it does not say definitively that he was intending suicide. Hence, there have been almost as many sightings of Majorana as of Elvis (at least among scientists). One almost creditable account is that he went to Venezuela, which is based upon an investigation by the Italian authorities this century, which was based upon analysis of an alleged photograph taken in Argentina in 1955 and a claimed meeting with him in Buenos Aires somewhat earlier. Coming back to the topic, there are recent experiments in nanowires adjacent to superconductors (so that an induced superconductivity in the nanowire gave access to the zero energy states) and in ferromagnetic atomic chains that are very suggestive of seeing the Majorana states. As a consequence, this has become a field of relatively high interest. Each such revelation of the possibility of new and different particles just highlights the obvious error of Heisenberg in claiming that quantum mechanics was complete. For this field, which continues to allow predictions of new and exciting possibilities for particles or other attributes, cannot be considered to have been complete.

Advance in Understanding

Also, in the year following the Solvay conference came another important interpretation. In May 1928, a paper from Kennard appeared in *Physical Review* [20], which reviewed the wave mechanics of

Schrödinger. He considered the motion of a series of particles moving quantum mechanically. While using the Schrödinger equation, he also adapted the quantum potential that arose from Madelung's earlier work [21]. From this, he reached the conclusion [20]:

> *Thus each element of the probability moves in the Cartesian space of each particle as that particle would move according to Newton's laws under the physical force plus a "quantum force".* . . .
>
> The motion here considered occurs in a space of n dimensions. We can also, however, replace the n-dimensional packet by n separate packets, one for each particle, all moving in the same ordinary space.

Kennard's quantum force was similar to the quantum potential due to Madelung, which provided a correction to induce quantum effects, such as interference. Because of the interaction between the particles, the motion of each particle depends upon the positions of all the other particles, and this provides a sense of nonlocal behavior in the motion. This nonlocality, at least in the many-particle problem, would be important in later arguments about the interpretation of quantum mechanics. Kennard [20] reached another important conclusion:

> In terms of these results the relationships between quantum and classical mechanics stand out very clearly. If we put $h = 0$, the probability becomes a distribution of matter moving classically. In general, *if a system is immersed in a uniform force field, its probability packet will simply execute the classical motion in that field in addition to whatever internal motion it may have* by reason of internal classical forces or the quantum force. . . .
>
> Thus the quantum mechanics predicts no novel interaction between electrons spaced so far apart that their classical interaction is slight.
>
> . . . *the center of probability for each particle moves in ordinary space as would the particle itself in classical mechanics under the "mean probable" force.*

This is a major connection between classical statistical mechanics and the quantum mechanics of the many-electron system. Fürth would reinforce this connection a couple of years later, when he made a formal connection between the motion of a classical statistical ensemble and quantum mechanics [22].

The connection between the quantum probability and that of a classical statistical system may seem a bit too much to hope for. However, it turns out that there is a strong connection, which goes beyond just the observations of these scientists. In classical statistical physics, the Fokker–Planck equation yields a probability distribution function for the statistical system [23]. This equation also allows for a *probability current* to describe the motion of the ensemble. The connection between the Schrödinger equation and the Fokker–Planck equation is strong. One can generally connect the probability current to an eigenvalue of the Schrödinger equation. One can actually create a reversible, and invertible, transformation which will take one equation into the other [24, 25]. This means that there is a real mathematical relationship between the solutions of one of these equations and the solutions of the other equation. Just as we observed earlier that quantization of the atom went away in four spatial dimensions, here we now have verification of the connection between classical statistical mechanics and quantum mechanics, and this occurs without any mumbo-jumbo hand waving.

Following the Solvay conference Schrödinger corresponded with Bohr with some questions about his paper on complementarity [26]. As he subsequently wrote to Einstein [27], he was having trouble with one of Bohr's assertions. Bohr had used the form

$$\int pdx = nh \rightarrow p_n = nh/2L, \qquad (5.5)$$

for quantization of a molecular motion, where L is the distance over which the molecule gets repeatedly reflected. Schrödinger pointed out that this made the various quantum levels so closely spaced that "even with the largest possible uncertainty in the *coordinate* ($\Delta x = L$), I cannot buy enough accuracy in the *momentum* to allow me to distinguish between neighboring quantum states." Einstein [28] responded:

> I think that you have hit the nail on the head. It is true that the evasion using the arbitrarily large domain of cyclic variables to limit the value of Δp is very ingenious. But an uncertainty relation

interpreted that way does not appear to be very illuminating. The thing was invented for free particles, and it fits that case in a natural way. Your claim that the concepts of p, x will have to be given up, if they can only claim such a "shaky" meaning, seems to me to be fully justified. The Heisenberg–Bohr tranquilizing philosophy—or religion?—is so delicately contrived that, for the time being, it provides a gentle pillow for the true believer from which he cannot be easily aroused.

Once more it appears that Bohr has made claims that go beyond the normal interpretations. Mathematically, the problem in question seems to be a typical eigenvalue problem, and today we do not associate any uncertainty to solutions of such a problem. But this would not have deterred Bohr from propagating his views and claims, as we will see.

Einstein–Bohr Once Again

Einstein and Bohr would face off once again at the 1930 Solvay conference. This time the topic was magnetism, which implied that there would be discussion about the role of quantum mechanics and electron spin, all of which contributed to magnetic phenomena. Einstein would of course be there as a member of the committee, but Bohr, Dirac, Heisenberg, Kramers, and Pauli would also be invited. De Broglie and Schrödinger, however, would not attend. So, other than Einstein, the field was essentially left to the acolytes of the Copenhagen doctrine.

Einstein had come up with another proposal for a "thought" experiment which would address the uncertainty principle, this time using an energy-time uncertainty. As remarked in the last chapter, however, it had been pointed out that energy-time was not a good quantum mechanical uncertainty relation in classical quantum mechanics. But in relativistic quantum mechanics, this is no longer the case, as Einstein suggested.[b] His idea was to consider a box which contained a reasonable number of photons. He would also

---

[b]In classical mechanics and normal quantum mechanics, time is a parameter which characterizes the forward progress of the system. In the relativistic world, however,

place a clock inside the box, so that a shutter would be opened and closed at specific times, thus allowing a single photon to escape from the box. In addition, the box would be suspended from a very accurate scale. Einstein's proposal was that, since the energy of the photon that left the box would change the weight of the box through Einstein's famous $E = mc^2$, this weight change could be recorded by the scale. In addition, the time during which the shutter was open would be known very accurately. Hence, these two measurements taken together would exactly determine the change in energy and time. This proposal apparently was presented as just discussion between Einstein and Bohr in the hallway [29, 30]. Nevertheless, the audience who watched the debate certainly was large. Bohr was basically stunned. As usual, he did not know how to reply, and he spent the rest of the day discussing it with his acolytes.

By the following morning, Bohr had his reply, which in essence, was composed of three steps [31]. The first step sets up an uncertainty in the momentum. If the energy of the box is determined with a precision ΔE, then this leads to an uncertainty in the mass through Einstein's equation cited above, and this in turn leads to an uncertainty in the gravitational force acting upon the box. Since the time for which the shutter is open is known, the uncertainty in the force leads to an uncertainty in the momentum transferred to the box by the departing photon. So far, this does not contradict Einstein's claim to measure the energy change with arbitrary precision. Now, Bohr points out that relativity theory couples time and position with the gravitational force. That is, this leads to the problem that the position of the box affects the clock due to the slight change in gravitational force as the box moves from one position to another. Therefore, there is an uncertainty in the time of the photon emission due to the movement of the box. The third step is to relate the uncertainty in energy to the uncertainty in the time of emission, which yields an uncertainty between the momentum transfer and the position of the box, which satisfies the Heisenberg uncertainty relation.

---

space variables and time are on an equal footing, forming the four-dimensional world. Hence, in this latter case, one can have an energy-time uncertainty.

Bohr had used Einstein's own general theory of relativity to show that his proposal must obey the Heisenberg uncertainty relation. Contrary to the discussion at the 1927 conference, Bohr allegedly has now shown, for the first time, that Einstein's proposals are subject to quantum uncertainty rather than merely classical uncertainty, up to a point. The point is that Bohr did not need the extra factor of 2 discussed in the previous chapter. Rather, he showed only that the energy-time uncertainty produced a momentum-position uncertainty. He did not show that it had to be greater than $\hbar/2$. So, while he claimed that the proposed measurements must be limited by *some* uncertainty principle, he did not demonstrate that this had to be the quantum uncertainty. Bohr's entire argument was classical even though it entailed general relativity. The step to quantum uncertainty was a claim, which was unproven. Yet, the general observers felt that Bohr had won a great victory. Of course, with victory come the spoils, and in this case, it was the pronouncement that quantum mechanics must be consistent across the many various arguments. Some have argued that, in fact, it was not Bohr who introduced relativity, but Einstein himself when he used his energy–mass equation. There is also the concern that Bohr's "proof" demonstrated that quantum mechanics is consistent with general relativity but not with Newton's theory of gravity [30]. Does this put too much value on the state of quantum mechanics? Worse, there is the overriding view of the Copenhagen interpretation that was supported by Bohr's "victory," in that the escaping photon was produced by the measurement itself; it was not defined until the moment of measurement. If one does not measure the photon, it is nowhere [29].

So, an *ex post facto* examination of the problem and its solution does not really support any claim that Bohr had won the argument. He had certainly shown that an uncertainty existed in the problem, which prevented Einstein from measuring time and energy exactly. But he had not shown that this required quantum uncertainty, as opposed to normal classical uncertainty. For the latter, we have already shown in a previous chapter that classical arguments certainly lead to a classical uncertainty, which is not much different from that proposed by Heisenberg. Hence, the claim that Bohr had shown that quantum mechanics was consistent and correct is just not supported by the evidence. This result could be claimed only if one also supposed that there was no uncertainty in classical measurements. Not even Bohr could claim this was the case.

Moreover, Einstein now turned to worry about whether quantum mechanics could be complete. That is, was the theory complete? The obvious answer was "no," simply because the Dirac equation had arrived the previous year with the prediction of new anti-particles (generalizing from the exact prediction of the positron). But, as Pais [16] said,

> It was his [Einstein's] almost solitary conviction that quantum mechanics is logically consistent, but that it is an incomplete manifestation of an underlying theory in which an objectively real description is possible—a position he maintained until his death.

If we are to accept the work of Kennard and Fürth mentioned above, then Einstein was not alone in this conviction. There is also the feeling that Einstein's next challenge actually grew from the problem he posed at this 1930 conference, but that challenge would have to wait, as political events were overtaking science.

For his part, Bohr turned more toward philosophy. He was so taken with his concept of complementarity for quantum mechanics, he was convinced that it would be applicable to almost every field of science, as well as to general philosophy. So, he was sure that the philosopher would rush to accept his great insight. It was not to be. As we pointed out in Chapter 3 [32],

> we realize that Bohr's proposal of the complementarity principle was nothing less than an attempt to make it the cornerstone of a new epistemology.... Whatever the most prominent factors were which contributed to Bohr's formulation of the complementarity point of view in physics—whether his physical research or thoughts on psychology, or reading in philosophical problems, or controversy between rival schools in biology, or the complementary demands of love and justice in everyday dealings—it was the *universal* significance of the role of complementarity which Bohr came to emphasize.
>
> Bohr's aim has a grandeur which one must admire. But while his point of view is accepted by the large majority in physics itself, it would not be accurate to say that it is being widely understood and used in other fields; still less has it swept over philosophy.

Perhaps the philosophers just didn't understand him. Perhaps they were not as accepting of the shallow thought as his acolytes [33]:

> His disciples transformed what would have been unaccept-able in any other physicist—his inability to do the necessary mathematics—into a virtue.... Bohr's unintelligibility as he struggled to explain his ideas was interpreted as a product of the great suffering he endured in his quest for understanding.

Beller [34] puts it in a similar manner:

> Elevating Bohr's depreciation of mathematics into an admirable virtue, Bohr's devotees believed that, unlike ordinary mortals, Bohr did not need to calculate in order to obtain the "truth."
> Bohr and his faithful disciples ... complained that philoso-phers and other nonscientific intellectuals did not comprehend the subtlety, profundity, and objectivity of complementarity.
> ... the more incomprehensible Bohr's words became, the more his devotees' conviction grew that a great truth, inaccessible to simple mortals, was hidden therein. They would spend hours, sometimes years, trying to recover the hidden meaning.

It seems that the rest of the world, other than his circle of devotees in physics, took him to be precisely as John Slater described after a very short time in Copenhagen (see the quote in Chapter 3).

The Acolytes' Reach

The idea that there was more to the atom than just the protons and the electrons was apparent almost from the beginning, even though there was some questions about what the extra bits were, but one has to realize that the community was just coming to grips with the idea that the atom was not the smallest particle in nature. From about the time of Moseley's experiments (just before WW1), ordering in the periodic table was according to the atomic *number* of the atoms, and not according to the atomic *weight*. In every case the atomic weight was larger, and sometimes much larger, than the atomic number. So, to what was this added weight due? Rutherford summarized the situation in 1920 by asserting that the nucleus contained both protons and electrons, with the atomic number being the difference between the total number of protons and the number of electrons in the nucleus [35], but he posited:

> The question whether the atomic number of an element is the actual measure of its nuclear charge is a matter of such fundamental importance that all methods of attack should be followed up.

So, we see that, while he asserted that there were electrons in the nucleus, he did not rule out the possibility that the extra weight (in excess of the atomic number) was not actually due to extra protons and electrons in the nucleus. He also used this argument to explain isotopes of an element as having different excess protons in the nucleus. Nevertheless, he thought,

> In the case of the nucleus, the electron forms a very close and powerful combination with the positively charged units and, as far as we know, there is a region just outside the nucleus where no electron is in stable equilibrium.

His support for such a view came from the fact that beta emission from the nucleus consisted of electrons! He would go on to call these bound protons and electrons neutrons. In the early 1930s, it was found that an unusually effective (at penetrating other material) radiation was produced from some atoms when bombarded by alpha particles. James Chadwick, working at the Cavendish laboratory in Cambridge, showed that these particles were neutral particles with about the same mass as the proton [36]. If this new particle were from a bound electron and proton from the nucleus, one would expect them to dissociate after leaving the nucleus. The fact that it remained neutral implied a new particle. Chadwick would win the Nobel prize for his discovery in 1935. In the meantime, his student Maurice Goldhaber would measure the mass very accurately [37]. So, the atomic world had a new particle, but more would come.

In the decay of beta radiation within the nucleus, beta radiation is emitted, but energy, momentum, and angular momentum (in particular, the spin angular momentum) were not conserved. Bohr, of course, said this was just what one might expect if the Bohr–

Kramers–Slater paper was correct. Pauli took an opposite view, and suggested that there was another explanation. His viewpoint became widely known, but did not appear in print until the proceedings of the seventh Solvay conference, where he discussed this proposal during the discussion session [38]. His idea was that the beta radiation was accompanied by a new particle, which was also neutral, but with an extremely small mass. And it possessed a spin, making it a fermion. While he wanted to call this very light particle a neutron, the name had already been taken. The final name of neutrino was due to Enrico Fermi, and Pauli used it at the Solvay conference. Fermi's theory of beta decay decreed that the neutron could itself decay into a proton, an electron, and this new particle— the neutrino. His paper was rejected by a British journal for being too far from reality, so was published in an obscure Italian journal and then in German [39]. It would be another two decades before the neutrino would be experimentally observed. But the world now had another new particle, and this one demonstrated that there were to be particles much smaller than even the electron. There is still debate about whether the neutrino has mass—if it had mass, it could not travel at the speed of light which it supposedly did. Modern experiments seem to support this speed, but the debate is far from over. There is even speculation that the neutrino might be its own anti-particle, which would mean it is the long searched for Majorana fermion. Pauli, for his part, would move to soft matter physics, concentrating on the theory of electrolytes, proteins, and colloidal solids. Heisenberg was intrigued by the more complex structure of the nucleus and would focus his attention there. This attention, and that of others, opened the field of nuclear physics.

But Pauli had opened a can of worms for Bohr. Pauli was an acolyte, yet here he had predicted a new particle, which was required for the conservation laws. Moreover, Fermi had described its properties through his theory. Thus, this new particle had well defined properties, and yet there was no measurement to experimentally confirm its existence. Are we to believe that this particle was real and possessed its own concept of reality without any observations? This was a step too far for Bohr, and it certainly challenged his Machian view of reality. Moreover, this new particle, as well as Dirac's positron, would certainly fit into Einstein's version of reality, and this would become quite apparent just a few years later. The world, and quantum mechanics, was not going quite the way Bohr envisioned it.

Paul Dirac would publish his take on quantum mechanics in 1930 [40]. Born also published his new book, written with Pascual Jordan, on quantum theory in 1930 [41]. But perhaps the most influential book was published by John von Neumann in 1932 [42]. Von Neumann had been a student in Göttingen, and worked on the Hilbert space, a linear algebra formulation of quantum mechanics that would be peculiar to matrix mechanics. But it also allowed connection to wave mechanics. The book was a combination of his earlier papers on the subject and a serious investigation of both causality and the measurement problem in quantum mechanics. Here, he studied the ensembles of discrete systems to further the statistical approach. One of his most important considerations was as to whether the statistical approach could be considered to be closed or whether another deterministic approach with hidden variables was possible. He decided that hidden variables were not possible, and that the statistical approach was complete. It was subsequently shown that [43]

> it was not the objective measurable predictions of quantum mechanics which ruled out hidden variables. It was the arbitrary assumption [by von Neumann] of a particular (and impossible) relation between the results of incompatible measurements either of which *might* be made on a given occasion but only one of which can in fact be made.

Cushing also suggested that the von Neumann "theorem" was too restrictive in its reach. Instead, what is really required is [44]

> that such a hidden variables extension requires a *more* complete specification of the state of a microsystem than that possible with the quantum-mechanical state vector ψ.

In essence, von Neumann had taken a step too far. While the mathematics looked good, the underlying physics that he considered was not up to the task.

Von Neumann introduced another important process in the book, and this is one that had been discussed by the Göttingen crowd for some years. This was the so-called "wave function collapse." When the quantum system interacted with the classical measuring system, it required that the quantum system be projected into one of the classical states of the measurement system. Thus, the quantum probability would transition into a classical certainty. This was an essential part of the philosophy of Bohr and the acolytes. It meant that once the measurement was made, there was no way to make a second measurement on the quantum properties that existed prior to the first measurement. While a wave function for the entire quantum system was supposed to exist, the measurement would entangle it with the measuring system, with the end result that projected everything into one of the possible states of the measuring system. The wave function was no longer there! Again, this would be a major point in the coming challenge of Einstein to Bohr. But was it correct? There certainly was some debate about this point, and the more modern theory of weak measurements yield approaches for which the wave function is assumed not to (at least completely) collapse. We return to this in the later chapters.

The Exodus

On January 30, 1933, Adolf Hitler was sworn in as the new chancellor of Germany. On February 27, the Reichstag (German Parliament) building burned, under very questionable circumstances, and became unusable. On March 23, the remaining parliament was assembled in a tense environment and passed the "enabling act." This act allowed Hitler to pass laws without consent of the parliament for four years. Democracy was gone in Germany. On April 7, Hitler decreed an important new law. This was called the "Law for the Restoration of the Professional Civil Service," yet it was effectively the law to remove all people of Jewish origin, and all political opponents of Hitler, from their government positions. Within a few weeks, all Jewish professors in Göttingen, Berlin, and elsewhere, were suspended from their positions. This included Max Born and Albert Einstein. While Schrödinger was not Jewish, he was known to be an ardent anti-Hitlerite, so he held out no hope for his own survival under the law. Pauli was in Zürich, so was safe from Hitler, although not from his own alcoholism. Heisenberg was not Jewish, and he would become important within Germany and lead their development efforts for the "bomb." Jordan was already a supporter of Hitler, so was also allowed to become influential. Kramers was in Belgium and Bohr in Denmark, so both were out of Hitler's reach, at least for the time being. The theoretical centers in Munich and Göttingen were gone, and the experimental centers in places like Berlin would soon follow. Most of the players were now more concerned with trying to find a place to survive, rather than pushing the frontiers of science. Only Bohr remained secure, free to promulgate his philosophy, basically without opposition. Born described the situation admirably, in a letter to Einstein [45]:

> I wish I could help you to look after the young exiled physicists . . ., but I am in the same position myself. . . . I do have, after all, one advantage: . . . I have plenty of time at my disposal. But it is not too easy without a library.
>
> I have never felt particularly Jewish. Now, of course, I am extremely conscious of it, not only because we are considered to be so, but because oppression and injustice provoke me to anger and resistance.
>
> I would like my children to become citizens of a Western country, preferably England, for the English seem to be accepting the refugees most nobly and generously.

In early 1933, Einstein was visiting the United States. With the events in Germany, he knew he could not return. To Hitler, he was the most Jewish of the scientists, and it was specifically Einstein's Jewish science that Hitler wanted to eradicate. At the end of March, Einstein returned to Europe, although he remained in Belgium. He revoked his German citizenship and turned in his German passport. That summer, he was invited to visit Britain and, while there, had the opportunity to meet with Churchill and other political leaders. Here, he urged them to accept the refugee scientists, which helped to open Britain to them. Finally, later that year, he returned to the United States to take a position at the Institute for Advanced Study at Princeton. It would be another two years before he finally decided to make this his permanent home. He was followed to Princeton, and the IAS, by John von Neumann, also in 1933, and Kurt Gödel, the famous Austrian mathematician and philosopher, in 1940. All would remain at IAS throughout their career.

During the early summer of 1933, the Borns and Schrödingers had both gone to northern Italy for vacation. Schrödinger's assistant, Arthur March, and March's wife Hilde had joined them for the vacation, as Erwin thought that he was in love with Hilde. This was merely a lull before the transition. Heide Born was getting the family ready to leave Germany. Born would take up a position at Cambridge that fall, at St. John's College, but this was a only temporary position. In 1935, his position at Göttingen was officially terminated and his German citizenship was revoked. As his position in Cambridge was only temporary, he considered several offers, until he was finally given the chair in natural philosophy at Edinburgh University. For Born, he commented in a letter he had written to Einstein [46]:

> My appointment to the University of Edinburgh is mentioned here for the first time. To us it meant the end of uncertainty, and the beginning of a new life in Scotland.

Born became a British citizen and remained at Edinburgh until his retirement at age 70. Schrödinger arranged for a position for himself and his assistant at Oxford University. However, the locals did not particularly appreciate his lifestyle, and he returned to Austria in 1936 to a position in Graz. In 1938, after the Anschluss, this position also became untenable for him. Fortuitously, the president of Ireland wanted to establish an Institute for Advanced Studies in Dublin, and offered Schrödinger the first official position for the new entity. He moved to Dublin and became the Director of the School for Theoretical Physics. Leaving Austria was difficult and quite complicated, although it was finally accomplished. Schrödinger was quite comfortable finally in Ireland. As Anny [47] said,

> We were very, very happy when we at last arrived in Dublin. It was seventeen years we stayed there. It was wonderful, peaceful. Really, he liked it very, very much.

> It was especially nice in Dublin ... because he had no duties whatsoever. The whole scientific [life] coming together. The main thing was to teach in the morning and to hold a colloquium or seminar in the afternoon. Never real lectures but always a conversation. He had a very nice room and they [the students] could come and speak to him whenever they liked. They appreciated it very much.
>
> In Dublin he went everyday to the institute. He went every day, either in the noon or in the afternoon....
>
> Thirring [another Austrian physicist] was there too. He was very young; just after he had made his doctorate he came to Dublin. There were about ten or twelve students and lots of nationalities. They were very international. Of course, Ireland was not in the war, you know, and everybody loved to come there.

More remarkable was that the Irish tolerated the lifestyle that Schrödinger loved to pursue. With his wife and a mistress (Hilde March), and the daughter Ruth, from the latter, he would settle in Clontarf, a suburb of Dublin [48]. They had a normal house, typical of the middle class, which they first rented, and then managed to buy. He is thought to have fathered two other daughters, by two separate women, during his time in Dublin [49]. This would be in keeping with his alternate passion—L'Amour. He would remain in Dublin until 1955, when he returned to Vienna to take up a chair in physics at the University.

During a trip to the United States in 1931, Pauli was supposed to lecture at the University of Michigan. Unhappy with prohibition, he slipped into Canada, where a drinking bout led him to fall and break his shoulder [50]. His drinking problem would become worse in Zürich. Finally, an intervention by his father led him to seek help with the problem and he underwent psychoanalysis. With the Anschluss, Wolfgang Pauli became a German citizen, and this created problems for him. Because he was a professor in Zürich, he tried to obtain Swiss citizenship. For a variety of reasons, this came to nothing. Finally, in 1940, he joined Einstein at the Institute for Advanced Study at Princeton, becoming a US citizen in 1946. He finally was granted Swiss citizenship in 1949, where he had returned after the war.

In Germany, there would arise a movement, which has been termed *Deutsche Physik*. It was extremely anti-Semitic and very antitheoretical physics. In their minds, theoretical physics was Jewish physics. It seems that the return to Machism was very convenient for the Nazi party. Yet, Bohr's insistence on measurement and experiment provided a philosophy which fit right in to this new movement. The movement became a problem for Heisenberg, as a theorist, when he was nominated for a professorship in Munich just before the war. His selection was finally pushed forward by party officials, for a variety of reasons. One was the recent discovery of nuclear fission. The government had begun the nuclear weapon project, and Heisenberg became one of the leaders in research and development for this project. He would travel to Copenhagen, but Bohr was not comfortable with his effort on this project, particularly after the Germans occupied Denmark. During the war, Heisenberg became professor at Humboldt University in Berlin (at that time, it was called the Friedrich Wilhelm University). Fortunately, the nuclear effort was largely futile. Heisenberg would be interned by the British after the war, which is another saga.

References

1. G. Bacciagaluppi and A. Valentini, *Quantum Theory at the Crossroads* (Cambridge Univ. Press, Cambridge, 2009).
2. P. A. M. Dirac, *Proc. R. Soc. London, A* **117**, 610 (1928).
3. E. Schrödinger, *Ann. Phys.* **81**, 109 (1926).
4. W. Gordon, *Z. Phys.* **40**, 117 (1926).
5. O. Klein, *Z. Phys.* **41**, 407 (1927).
6. W. Pauli, *Z. Phys.* **43**, 601 (1927).
7. P. A. M. Dirac, *Proc. R. Soc. London, A* **118**, 351 (1928).
8. D. R. Hartree, *Proc. Camb. Phil. Soc.* **24**, 89 (1928)
9. W. Heisenberg and W. Pauli, *Z. Phys.* **56**, 1 (1929); **56**, 168 (129).
10. P. A. M. Dirac, *Proc. R. Soc. London, A* **136**, 453 (1932).
11. P. A. M. Dirac, *Proc. R. Soc. London, A* **126**, 360 (1930).
12. H. Weyl, *Z. Phys.* **56**, 330 (1929).
13. P. A. M. Dirac, *Proc. R. Soc. London, A* **133**, 60 (1931).

14. P. A. M. Dirac, in American Institute of Physics Oral Histories: https://www.aip.org/history-programs/niels-bohr-library/oral-histories/4575-5

15. W. Heisenberg, in American Institute of Physics Oral Histories: https://www.aip.org/history-programs/niels-bohr-library/oral-histories/4661-10

16. A. Pais, *Niels Bohr's Times, in Physics, Philosophy, and Polity* (Clarendon Press, Oxford, 1991), p. 356.

17. See, e.g., F. Close, *Antimatter* (Oxford Univ. Press, Oxford, 2009).

18. C. D. Anderson, *Phys. Rev.* **43**, 491 (1933).

19. E. Majorana and L. Maiani, *Il Nuovo Cim.* **14**, 171 (1937).

20. E. H. Kennard, *Phys. Rev.* **31**, 876 (1928).

21. E. Madelung, *Z. Phys.* **40**, 322 (1926).

22. R. Fürth, *Z. Phys.* **81**, 143 (1933).

23. H. Risken, *The Fokker-Planck Equation* (Springer, New York, 1984).

24. A. Schulze-Halberg and J. M. C. Jimenez, *Symmetry* **1**, 115 (2009).

25. D. K. Ferry, *J. Comp. Theor. Nanosci.* **8**, 953 (2011), and references therein.

26. N. Bohr, *Naturwissen.* **16**, 245 (1928).

27. E. Schrödinger, letter to Einstein of May 30, 1928; reprinted in K. Przibaum, *Letters on Wave Mechanics* (Philosphical Library, New York, 1967).

28. A. Einstein, letter to Schrödinger of May 31, 1928; reprinted in K. Przibaum, *Letters on Wave Mechanics* (Philosphical Library, New York, 1967).

29. L. Blom, *On the Uncertainty Principle: The Quantum Debate Between Bohr and Einstein*, Bachelor's thesis, University of Amsterdam, 2009. http://dare.uva.nl/document/169878

30. D. Home and A. Whitaker, *Einstein's Struggles with Quantum theory: A Reappraisal* (Springer, New York, 2007).

31. H. Nikolic, *Euro. J. Phys.* **33**, 1089 (2012).

32. G. Holton, *Thematic Origins of Scientific Thought*, 2nd ed. (Harvard Univ. Press, Cambridge, 1988), pp. 134–138.

33. S. Jones, *The Quantum Ten* (Thomas Allen, Toronto, 2008), p. 278.

34. M. Beller, *Osiris* 14, 252 (1999).

35. E. Rutherford, *Proc. R. Soc. London, A* **97**, 374 (1920).

36. J. Chadwick, *Nature* **129**, 312 (1932).

37. J. Chadwick and M. Goldhaber, *Nature* **134**, 237 (1934).

38. W. Pauli, in *Rapports du Septième conseil de Physique Solvay* (Gauthier-Villars, Paris, 1934), p. 324.

39. E. Fermi, *Z. Phys.* **88**, 161 (1934).

40. P. A. M. Dirac, *The Principles of Quantum Mechanics* (Clarendon Press, Oxford, 1930).

41. M. Born and P. Jordan, *Elementare Quantenmechanik* (Springer, Berlin, 1930).

42. J. von Neumann, *Mathematische Grundlagen der Quantenmechanik* (Springer, Berlin, 1932).

43. J. Bell, *Rev. Mod. Phys.* **38**, 447 (1966).

44. J. T. Cushing, *Quantum Mechanics: Historical Contingency and the Copenhagen Hegemony* (Univ. Chicago Press, Chicago, 1994), p. 134.

45. M. Born, letter to A. Einstein of June 2, 1933, reprinted in *The Born-Einstein Letters*, trans. by I. Born (Macmillan, London, 1971).

46. M. Born, in Ref. 45, following his letter to Einstein of August 24, 1936.

47. A. Schrödinger, in American Institute of Physics Oral Histories: https://www.aip.org/history-programs/niels-bohr-library/oral-histories/4865

48. J. Gribbon, *Erwin Schrödinger and the Quantum Revolution* (Wiley, Hoboken, 2013).

49. R. Fanning, *Eamon de Valera: A Will to Power* (Harvard Univ. Press, Cambridge, 2016).

50. S. Jones, *op. cit.*, p. 275.

Chapter 6

1935

While the period following the 1927 and 1930 Solvay conferences seemed to be relatively calm, a series of important papers would arrive in 1935. These included Schrödinger's view on the status of quantum mechanics as well as his famous thought experiment involving a cat. In addition, there came from Einstein his final attack on the completeness of quantum mechanics, and Bohr's supposedly thorough defeat of the idea. As we will see, one's view of the entire collection of papers and their impact really depends upon whether or not one is a devotee of the Copenhagen ideas. Together, these papers would challenge the philosophical basis of quantum mechanics and bring the various issues to the front. Once again, however, the so-called community of physicists would decide that Bohr successfully defended his interpretation. Perhaps this is only an overreaction, or a predisposed belief that the religion must be defended against all agnostics who fail to see the rational thought that supports the religion. This latter view, though, jumps to the conclusion that such a rational thought process underlies the religion. If we are to believe Slater's and Gellman's comments (Chapter 3), this just may not be the case.

The Copenhagen Conspiracy
David Ferry
Copyright © 2019 Pan Stanford Publishing Pte. Ltd.
ISBN 978-981-4774-75-8 (Hardcover), 978-1-351-20723-2 (eBook)
www.panstanford.com

Schrödinger

During 1934–5, Erwin Schrödinger was primarily situated at Oxford, although he toured the United States during this period. During 1935, he received support from the Imperial Chemical Industries, London, which allowed him some leisure time in which to write an overview of the state of quantum mechanics as he saw it. Mainly, however, he attacked the logic of the Copenhagen–Göttingen orthodoxy. The article actually appeared after the Einstein–Podolsky–Rosen (EPR) paper, but we want to discuss it first, as it thoroughly goes through a great many discussions on the philosophical basis of quantum mechanics. Schrödinger's work would be published in three parts in the German journal *Die Naturwissenschaften* (The Science of Nature) [1]. We have become more familiar with the work through the translation by J. D. Trimmer [2]. One of the essential elements which he put forward was the idea of *entanglement*, the connection between separated systems. He would expand upon this in two additional articles [3].

Schrödinger begins by describing the classical system, using the Rutherford model of the atom. Here, one has a single nucleus and a single electron, each of which has position and momentum. Thus, in our three-dimensional world, this leads to 12 components or variables. There is only one system, but there are several ways in which the 12 variables can be defined (e.g., absolute coordinates with respect to the observer's world or center-of-mass coordinates are two such examples). The connection between classical mechanics and quantum mechanics has been described by Trimmer [2]:

> It [quantum mechanics] takes over unquestioningly from the classical model the entire infinite roll call of imaginable variables or determining parts and proclaims each part to be *directly measurable*, indeed measurable to arbitrary precision, so far as it alone is concerned.
>
> The classical concept of *state* becomes lost, in that at most a well-chosen *half* of a complete set of variables, can be assigned definite numerical values.... The other half then remains completely indeterminate.

One must guard against criticizing this doctrine because it is so difficult to express.... But a different criticism surfaces. Scarcely a single physicist of the classical era would have dared to believe, in thinking about a model, that its determining parts are measurable on the natural object. Only much remoter consequences of the picture were actually open to experimental test.... while the new theory calls the classical model incapable of specifying all details of the *mutual interrelationship of the determining parts* (for which its creators intended it), it nevertheless considers the model suitable for guiding us as to just which measurements can in principle be made on the relevant natural object. This would have seemed to those who thought up the picture a scandalous extension of their thought-pattern and an unscrupulous proscription against future development.

Here, then we have the direct challenge to Bohr and his philosophy. Schrödinger suggests that the Bohr doctrine is "born of distress." He then discusses the failures of the statistical approach. For example, one cannot meaningfully discuss the probability of finding an energy level of an oscillator between E and E' if the separation of these two energies is less than the known energy level spacing of the oscillator. Of course, classical distributions represent the results of many measurements made on an ensemble of systems as a means of building up some idea of the possible results of any single measurement. To give exact values to an ensemble of classical states would conflict with the ideas of quantum mechanics with its uncertainty relations. In some sense, this can be viewed as a blurring of the values of the various parameters or variables. Yet, he remarks [2],

> That it is not impossible to express the degree and kind of blurring of *all* variables in *one* perfectly *clear* concept follows at once from the fact that Q.M. as a matter of fact has and uses such an instrument, the so-called wave function....
>
> In fact the function has provided quite intuitive and convenient ideas, for instance the "cloud of negative electricity" around the nucleus.... But serious misgivings arise if one notices that the uncertainty affects macroscopically tangible and visible effects, for which the term "blurring" seems simply wrong. The state of a radioactive nucleus is presumably blurred in such degree and

fashion that neither the instant of decay nor the direction, in which the emitted α-particle leaves the nucleus, is well established.

Yet, *a measurement of the emitted particle shows that it arrives at the screen at a particular point, not as a blurred-out spherical wave.* Blurring seems actually to be the wrong interpretation, since the well-known cloud chamber tracks are anything but blurred.

Schrödinger's Cat

As an example of a "ridiculous" case, Schrödinger then proposes his cat experiment. Here, he puts a cat into a steel enclosure, such as a box. Inside the box, he places a Geiger counter containing a small amount of a radioactive substance. The amount is sufficiently small that within a given period (he proposes an hour) one of the atoms decays with a probability of one-half. If the decay occurs, the counter will signal a relay to shatter a flask of hydrocyanic acid that is also within the enclosure. If the decay occurred within this given period, the cat is dead, poisoned by the release of the acid. If no decay occurred, the cat lives. The wave function for the entire system must express these two results with equal probability. Here, the conundrum is that a probability event that is originally limited to the atomic nucleus has now been transferred to the entire macroscopic system. That is, the cat is a classical object. It eats, purrs, defecates, and sharpens its claws on the curtains or the furniture. There is nothing quantum mechanical about it (except perhaps the fundamentals of its biological functions). Similarly, the box, the counter, and the flask are classical macroscopic objects. The idea of blurring is not appropriate to these variables or objects. How did the quantum mechanics get to the macroscopic level?

The question begins to appear as the philosophical basis upon which one should proceed. Schrödinger puts it simply [2]:

> We are told that no distinction is to be made between the state of a natural object and what I know about it, or perhaps better, what I can know about it if I go to some trouble. Actually—so they say— there is intrinsically only awareness, observation, measurement.

If through them I have procured at a given moment the best knowledge of the state of the physical object that is possibly attainable in accord with natural laws, then I can turn aside as *meaningless* any further questioning.

From this, we can infer that the quantum mechanics of our atom made the jump to the macroscopic system through the supreme arbiter of truth—the measurement. To achieve this, even the cat had to be given a quantum mechanical existence in order for the quantum system to extend to the measurement—the opening of the box by an observer. At the moment of truth, then, the quantum state must collapse into one of the possible results of measurement—dead cat or alive. But Schrödinger finesses the actual possible outcomes of the experiment, saying only [2],

The first atomic decay would have poisoned it [the cat]. The ψ-function of the entire system would express this by having in it the living and the dead cat (pardon the expression) mixed or smeared out in equal parts.

... would our measurement reality become highly dependent on the diligence or laziness of the experimenter, how much trouble he takes to inform himself [of the state of the natural object].

He left it somewhat vague in order to allow the reader to draw his own conclusions as to how to put together the wave function. The naïve reader would assume that only two possible solutions need be included—live cat and intact vial or dead cat and broken vial. Indeed, one can find many YouTube videos with just this result and discussion, and usually these pronounce victory for Bohr's measurement basis for quantum mechanics. This, however, does not exhaust the possible outcomes. Consider that a decay occurs, but either the counter does not register it or the vial survives the hammer blow, leaving the cat alive. Any reasonably competent experimentalist is familiar with this possible outcome, ascribing it to Murphy's law. For quantum mechanics, it would be an example of the Pauli effect [4],

the mysterious, anecdotal failure of technical equipment in the presence of ... Wolfgang Pauli.

Nevertheless, this still does not complete the possible outcomes of the experiment, which is why it is termed a "paradox." Suppose the cat died of suffocation and the decay did not occur. Or, suppose the nucleus did not decay until the moment just before the box was opened, so that the mechanical delay before the vial was broken meant this latter would occur after the box was open. Could that actually give another result—live cat and broken vial (but perhaps with a dead observer, if the wind was in the right direction). It is known that the radioactive decay does not stop when the box is opened. Suppose the observer opened the box and the cat jumped out and ran away; if the decay then occurred we could still have live cat and dead observer. This creates a true Machian conundrum. With the tree in the forest, the question was whether a falling tree made a noise if no one was there to sense the acoustic waves produced by the tree. Here, we have to ask *whether a measurement has been made if the observer is unable to communicate the results*. There are still many other possible outcomes of the experiment which one may imagine. Does the assumption of only two results overly restrict the possible models, avoiding more complete models which could have predicted a greater variety of results? Does choice of the measuring apparatus define the quantum system? Schrödinger would later say, with respect to arguments about the EPR paper [3]:

> It is suggested that these conclusions, unavoidable within the present theory but repugnant to some physicists including the author, are caused by applying non-relativistic quantum mechanics beyond its legitimate range.

In fact, a reasonable person would likely object to the extension of the quantum theory of the nucleus to the entire universe, an extension required when all possible outcomes are considered. Hence, one is drawn toward the conclusion that Bohr's insistence upon the measurement defining the quantum system is untenable. For, if no measurement can be said to have been made (dead observer), then there would seem to be no reality for either the cat or the remaining apparatus, which contradicts the construction of the experiment.

If we are to believe that Bohr's theory of the defining measurement is to confine our understanding of quantum mechanics, then we are forced to accept Schrödinger's comment [2]:

> For each measurement one is required to ascribe to the ψ-function ... a characteristic, quite sudden change which *depends on the measurement result obtained*, and so cannot be *foreseen*; from which alone it is already quite clear that this ... change ... has nothing whatever in common with its orderly development *between* two measurements.... It is precisely this point that demands the break with naïve realism.
>
> For *this* reason one can *not* put the ψ-function directly in place of the model or of the physical thing, ... because in the realism point of view observation is a natural process like any other and cannot *per se* bring about an interruption of the orderly flow of events.

Thus, it is clear that Schrödinger was not a believer in the Bohr form of the epistemology in which Mach's empiricism ruled the world. We will return to his further questioning of Bohr's philosophy later.

Entanglement

It was always more or less assumed that when two bodies interacted quantum mechanically, they would be linked much as the correlation that builds up between two bodies that collided classically. It was Schrödinger that really put the concept together in terms of the wave function, and it was he who coined the term *entanglement* to explain the interaction. As he says [2],

> If two separated bodies, each by itself known maximally, enter a situation in which they influence each other, and separate again, then there occurs regularly that which I have just called entanglement of our knowledge of the two bodies.

Generally, it is possible to know maximal information about each of the two bodies. If there is no interaction, the two bodies have no way in which to "learn" information about the other. A measurement of one can furnish no information about the other. The important thing about the entanglement is that when they interact, the two bodies are formed into a single system in which each body leaves traces of itself on the other body. After the interaction [2],

> the knowledge remains maximal, but at its end, if the two bodies have again separated, it is not again split into a logical sum of knowledges [*sic*] about the individual bodies. What remains of *that* may have become less than maximal, even very strongly so."

The important point to be made here is that, even after the two bodies have quit interacting and have moved some distance away from one another, they remain a single system, and the maximal knowledge of this entire single entity is likely more than the available knowledge of the original two bodies. Again, Schrödinger makes this point emphatically [2]:

> *Maximal knowledge of a total system does not necessarily include total knowledge of all its parts, not even when these are fully separated from each other and at the moment are not influencing each other at all.*

What happens in the simple case of two particles is that the interaction correlates the two positions and the two momenta. As a result, any measurement of one fixes the result for the other, and this means that we have lost some of the total possible knowledge during the interaction.

The single system of the two bodies remain so, until one or the other, or both, undergo interaction with another system. This may cause further entanglement or it may destroy the entanglement of the first system. The latter can be called a decoherence effect, after which the system again becomes two separate bodies. But we no longer have full knowledge of these two bodies, as we have not specified the nature of the secondary interaction, and the nature of the decoherence likely causes additional loss of information.

Schrödinger proposes a simple example of two systems with just a single degree of freedom in each, and we will follow through this example, paraphrasing his arguments. In this example, each of the systems has a single coordinate, which we can call x for the first particle and X for the second. To each of these coordinates there exists the canonically conjugate momentum. Again, we will call the momentum p for the first particle and P for the second. This particular simple system is important because it becomes (or became, as it was discussed earlier than the Schrödinger paper) the heart of the Einstein, Podolsky, and Rosen argument. If the two particles are allowed to interact, then entanglement occurs between them in the quantum mechanical case. It is convenient to organize the various coordinate systems (one for each particle, which do not have to be the same as they can be relative to some observer coordinate system), so that after the interaction, we have $x = X$ and $p = -P$. Now, this implies that a measurement of x on the first body yields a value which is exactly the same as a measurement of X on the second body, regardless of the order of these experiments. Similarly, we can say that a measurement of either of the momenta also determines both. A single measurement of position or momentum of the first body, or similarly of position or momentum of the second body gives us maximal information about the entangled system. However, there is an important caveat to all of this, as expressed by Schrödinger [2]:

> Every measurement is for its system the first. Measurements on separated systems cannot directly influence each other—that would be magic.

But a measurement of both position and momentum is not possible according to the accepted interpretation of quantum mechanics because there exists the uncertainty principle. Again [2],

> a prediction of this extent is thus utterly impossible according to the teaching of Q.M.[a] ... Many of my friends remain assured in this and declare: what answer a system *would have given* to the experimenter *if* ..., —has nothing to do with an actual measurement and so, from our epistemological standpoint, does not concern us.

---

[a] Here, one is referring to Bohr's picture of the philosophy of quantum mechanics.

"The quantum mechanician maintains that a [position]-measurement [means] ... my small system has a [wave]-function in which "*x* is fully sharp, but *p* fully indeterminate."

For sure, the measurement that is made must be the first, as we are told that the measurement collapses the wave function, so that no further measurements are meaningful on the system. Thus, we are forced to create multiple copies of the system, with a single measurement on each to form an ensemble of systems and measurements. The question may fairly be asked whether or not the empiricism implied in these measurements are really the governing concepts for quantum mechanics. Would a serious helping of realism change the picture? This is something we shall return to below.

Einstein, Podolsky, and Rosen

At the 1933 Solvay conference, the topic was on the structure and properties of the nucleus. Consequently, almost all the important figures in quantum mechanics were also there—Bohr, Schrödinger, Heisenberg, Pauli, Fermi, and both de Broglies. Surprisingly, Born was not present, although Kramers was there. Rosenfeld says that Einstein attended Bohr's lecture and listened intently. Then Rosenfeld says that, during the discussion, Einstein proposed [5]:

Suppose two particles are set in motion towards each other with the same, very large, momentum, and that they interact with each other for a very short time when they pass at known positions. Consider now an observer who gets hold of one of the particles, far away from the region of interaction, and measures its momentum; then, from the conditions of the experiment, he will obviously be able to deduce the momentum of the other particle. If, however, he chooses to measure the position of the first particle, he will be able to tell where the other particle is. This is a perfectly correct and straight-forward deduction from the principles of quantum mechanics; but is it not very paradoxical? How can the final state of the second particle be influenced by a measurement performed on the first, after all physical interaction has ceased between them?

Rosenfeld should have been an observer of this proposal, since he attended this conference. But Einstein did not. By October 1933, Einstein had only recently arrived in Princeton at the Institute for Advanced Studies, and does not appear in the official photograph of the conference (in fact, his absence is specifically noted). Later, in his review of the Solvay conferences, Bohr does not mention speaking at the 1933 conference [6], which was not in his nature to miss commenting on his own talks. (He remarks that he was invited to speak at the 1927 conference, which we have already seen was not the case.) While this discussion at the conference is perhaps wishful recollection on the part of Rosenfeld, the question raised is important.

The supposed problem is one of the key elements of the Einstein, Podolsky, and Rosen (EPR) paper in 1935 [7]. If Bohr had been asked the question earlier, why did he wait to respond until after the EPR paper appeared? Indeed, why does Rosenfeld then say about EPR that "this onslaught came down upon us as a bolt from the blue. Its effect on Bohr was remarkable" [5]? It seems to be inexplicable that a question asked in 1933 would be a shock in 1935. Yet a shock it would be to Copenhagen.

EPR begin by stating their philosophical view. It is an important establishment of how they feel the question of reality must be defined. Without this, the question stated above does not bring into focus the problem of whether quantum mechanics can be considered as a complete description of reality. And this latter is the question that they want to address. This philosophy is expressed as [7]:

Any serious consideration of a physical theory must take into account the distinction between the objective reality, which is independent of any theory, and the physical concepts with which the theory operates.

... we may ask ourselves two questions: (1) "Is the theory correct?" and (2) "Is the description given by the theory complete?" ... It is the second question which we want to consider here, as addressed to quantum mechanics.

... the following requirement for a complete theory seems to be a necessary one: *every element of the physical reality must have a counterpart in the physical theory.* We shall call this the condition of completeness.

> The elements of the physical reality cannot be determined by
> *a priori* philosophical considerations, but must be found by an
> appeal to the results of experiments and measurements. A ...
> definition of reality ... which we regard as reasonable [is] *If,*
> *without in any way disturbing a system, we can predict with*
> *certainty (i.e., with probability equal to unity) the value of a*
> *physical quantity, then there exists an element of physical reality*
> *corresponding to this physical quantity.*

We have to note that the authors are admitting that measurement
has a place in the understanding of the world, but insist that this
measurement produces a quantity that must have a counterpart in
the theory. That is, the theoretical description of the theory points
to the possible outcomes of the experiment, and the implication
is that this outcome would have the possibility of being measured
whether or not the measurement was actually made. Now, there
arises a philosophical consideration, referred to as counterfactual
definiteness. This is a classical concept which is deemed to not exist
in quantum mechanics. Two measurements of conjugate variables,
such as that of position and momentum, are excluded because they
cannot yield precise values for these two variables. Hence two
such measurements are said to be counterfactual. That is, they are
counter to the facts established by the interpretation of quantum
mechanics, in particular by the Copenhagen–Göttingen orthodoxy.
To say that our theory would allow us to consider measuring
the position of particle A or to *then* measure the momentum of
particle A, is explicitly excluded in quantum mechanics. To make
this suggestion, as is done in the example given and attributed
to Einstein by Rosenfeld, precludes the known interpretation of
quantum mechanics. The question then is where does reality and
completeness lie in quantum mechanics?

EPR then reviews some of the noncontroversial concepts of
quantum mechanics. If we consider that there is an operator O for
which the wave function satisfies an eigenvalue equation, then any
action of this operator on the wave function produces this particular
eigenvalue λ_O. In this situation, any particle in this state described
by this particular wave function leads to a physical reality for the
quantity described by O. On the other hand, any other operator,
which does not satisfy this eigenvalue expression also does not

have such an eigenvalue. Hence, for this second operator, there is no corresponding physical reality. When applied to the position and momentum, which are excluded from being simultaneously measured, EPR give the standard view [7]:

> When the momentum of a particle is known, its coordinate has no physical reality.
>
> From this it follows that either (1) *the quantum-mechanical description of reality given by the wave function is not complete* or (2) *when the operators corresponding to two physical quantities do not commute the two quantities cannot have simultaneous reality.*

As they then say, the normal situation in quantum mechanics is to assume that the wave function does, in fact, contain a complete description of the physical reality of the system in the state to which it actually corresponds. Now, this brings into focus the comments of Schrödinger, since if we describe the classical system by a quantum mechanical wave function, we must lose some information, because we can specify precisely only half as many variables. And EPR set out to demonstrate this contradiction.

For their approach, they in fact begin with the system outlined above, according to Rosenfeld. Two distinct systems are allowed to interact, and then are allowed to separate so that the interaction is no longer in force. But to pursue their argument through to the conclusion, we need to understand that they are proposing a total system with counterfactual definiteness. They acknowledge that the new wave function is such that they cannot determine the state of either of the two systems, but only their combined state. Determining the state of either of the systems requires measurements to be made. Now, they illustrate the problem. First, they consider measuring quantity A in system 1, in which A initially satisfies an eigenvalue equation. An important property of quantum mechanics, clearly spelled out in von Neumann's book [8], is that there is no preferred basis set of functions upon which to expand the wave function itself. One basis set can always be converted to any other by a similarity transformation. So, EPR now chose to expand the combined, two particle wave function in the eigenfunctions of operator A. Like a Fourier series, each term in the expansion is the product of an eigenfunction of system 1 for the operator A and a coefficient, which is a function of the variables of system 2. The basis functions from system 1 that are chosen of course depend upon the operator A that is selected. That is, the basis functions will be different, if a position operator is chosen, from those found for a different operator. Now, instead of operator A, they select a different operator B with its own set of eigenfunctions. The coefficients would now be different in the sum leading to the combined wave function. If they now measured either A or B, the remaining, collapsed wave function for system two would be different in the two cases. Hence, as a result of two different measurements on system 1, they find that system 2 is left in two different states and [7] *"it is possible to assign two different wave functions to the same reality."* They then consider an example in which the combined system is described as a plane wave-like wave function and show that it is possible to obtain the two wave functions for particle 2 by making either a position expansion or a momentum expansion. Hence, they have achieved for noncommuting operators simultaneous reality via the two separate wave functions for particle 2. From the two possibilities mentioned above, this then implies that the quantum mechanical description of physical reality given by the wave function is not complete.

In addition to completeness, they have also attacked the standard interpretation of the uncertainty principle. This had been Einstein's goal in both the 1927 and 1930 Solvay conferences. Bohr had shown that his constructions actually produced a *classical* uncertainty in the measurements, although he failed to establish that Heisenberg's uncertainty was fundamental in the arguments.

Von Neumann would argue, based upon the Göttingen mathematics as described in his book [8], that you cannot have two eigenvalues equations for noncommuting operators; e.g., two noncommuting operators cannot satisfy simultaneous eigenvalue equations. This would preclude the assumptions made by EPR, but this highlights the distinctions between the two understandings of quantum mechanics, because von Neumann would have insisted that the two measurements would not have been possible to make and that the two noncommuting operators could not possibly have simultaneous reality. Thus, quantum mechanics is complete in his world. Yet, EPR had done so. We are left with the conundrum, either the quantum mechanical wave function for two particles is not complete, or the two measurements cannot have simultaneous reality. To choose von Neumann's approach presupposes that the answer to the conundrum is known, and that the Copenhagen–Göttingen orthodoxy is the one true religion.

Bohr's Response

As was his wont, Bohr couldn't just respond with one paper. He had to first announce to the world that he was going to respond. This was done with a brief comment in nature [9]. Here, he pointed out that there was an apparent ambiguity in the problem posed in EPR:

> The procedure of measurements has an essential influence on the conditions on which the very definition of the physical quantities in question rests. Since these conditions must be considered as an inherent element of any phenomenon to which the term "physical reality" can be unambiguously applied, the conclusion of the above-mentioned authors would not appear to be justified.

In essence, he was retreating to his desired empiricism, by asserting that the measurement was important to the very definition of a physical quantity, which was a contrary view to that of either EPR or Schrödinger. What he is saying is that you cannot even define a physical quantity within quantum mechanics unless you describe the methodology of measuring such a quantity. The mere idea of physical reality could only occur after the measurement, and we are back to Mach's worldview. But Bohr also mentioned that a real reply would appear in the same journal as EPR.

In the longer reply [10], Bohr notes that EPR acknowledges that physical reality cannot be based solely upon philosophical arguments, but must be obtained from measurements. Then, he cites their statement about this:

> If, without in any way disturbing a system, we can predict with certainty the value of a physical quantity, then there exists an element of physical reality corresponding to this physical quantity.

Now, it must be noted here that "predict with certainty" has a different meaning than "measure." Bohr takes the opposite view [10]:

> "Such an argumentation ... would hardly seem suited to affect the soundness of quantum-mechanical description, which is based on a coherent mathematical formalism covering automatically any procedure of measurement like that indicated.

While EPR certainly described a mathematical procedure, it is not included in the above definition of physical reality. In addition, they discussed only the wave functions and never the procedure of measurement, but Bohr reiterates that the statement on reality has an essential ambiguity in his mind. He then wanders off to discuss other problems with which he can apply his theory of complementarity. These are essentially the problems raised by Einstein in the earlier Solvay conferences, so that he is restating his responses to Einstein's problems in somewhat more detail. His purpose in rehashing these problems lies, as he suggests [10],

> on the impossibility, in the field of quantum theory, of accurately controlling the reaction of the object on the measuring instruments, i.e., the transfer of momentum in the case of position measurements, and the displacement in case of momentum measurements.
>
> Indeed, in each experimental arrangement suited for the study of proper quantum phenomena not merely to do with an ignorance of the value of certain physical quantities, but with the impossibility of defining these quantities in an unambiguous way.

So, Bohr is demanding that the Copenhagen–Göttingen orthodoxy is correct and that it is impossible for EPR to do what they claimed to have done. *This avoids actually proving that they had made an error.* He continues to provide an idea for a measurement of the two particles discussed by EPR and, not surprisingly, finds that there must be an uncertainty. But, as before, he fails to describe whether this is normal classical uncertainty or is a proof of the necessity of Heisenberg's fundamental uncertainty. This argument does allow him to state [10],

> if we chose to measure the momentum of one of these particles, we lose through the uncontrollable displacement inevitable in such a measurement any possibility of deducing from the behavior of this particle the position of the diaphragm [through which the particle passed] relative to the rest of the apparatus, and have thus no basis whatever for predictions regarding the location of the other particle.

Bohr then returns to the clock problem from the 1930 conference, as a means to talk about the essence of time in the measurements, but this adds nothing to the essentials of the argument.

It is clear that Bohr is talking at cross-purposes to EPR. The latter discuss exactly the nature of the wave function and the existence of two possible states for particle 2, without actually making any measurement. Their treatment is mathematical using the expectation value of an operator. Connection to a real experiment lies only in the assumption that it would produce an eigenvalue of the operator. Bohr discusses the measurement problem, because in his mind, there is no possible reality outside of the measurement. Thus, for him, all is explained by his principle of complementarity. Now, von Neumann did not enter the fray at all, but had he done so, he might have pointed out that the measurement would have entangled the composite system and collapsed the wave function into *one of the possible eigenstates of the measuring apparatus*. In this case, there is no certainty that the eigenstates of the apparatus correspond at all to those of either operator A or operator B. So, for EPR, it is sufficient to discuss the results, *if* they had made a measurement on A or a measurement on B. In the EPR mind, the world of possibilities now includes the counterfactual definiteness which Bohr would refuse. But there remains an essential difference. EPR are arguing in terms of the eigenstates of the operators in the entangled system. Bohr would be arguing in terms of the eigenstates of the measurement apparatus. This difference is crucial to understanding that they have vastly different views of the epistemology of the system under discussion.

Others have also raised the questions about just what Bohr was trying to achieve in his response. Cushing points out [11]:

> Bohr's slip from epistemology (based on observability) to ontology (as a necessary discontinuity and as the impossibility of "classical" trajectories throughout the interaction) was ... not only logically unjustified but also not demanded, either by experiment or by the formalism of quantum mechanics.

Whitaker suggests that Bohr's view actually changed at this point in time, with his reply to EPR representing this change [12]:

> Unfortunately for Bohr, in 1935 Einstein, in collaboration with Boris Podolsky and Nathan Rosen, produced the famous EPR paper, which effectively demolished the disturbance interpretation. Bohr was forced to retreat towards conceptual argument. Indeed, his supporters could obtain sufficient justification to claim that the disturbance idea was never a central component of Bohr's ideas.

Here, disturbance is part of what has been referred to as collapse. The fact that the collapse was a central feature of the Göttingen mathematics is clear by the discussion in the von Neumann book. Yet, the push by Bohr's supporters was interpreted as saying that any criticism of the viewpoints must be based in a misunderstanding on the part of the critic. But the change became such that any account of the actual system under consideration now must include a complete description of the measuring apparatus as well as the observed results. This is certainly contrary to the view of EPR. Whitaker also points out that there are several well-known theories, such as thermodynamics, which are incomplete in the EPR sense. This was also apparent to the coming new generation of quantum physicists when Franco Selleri, in looking back at EPR, remarked [13],

> You can of course say that Einstein is wrong, anybody can be wrong, but you have to show that he is wrong. You cannot say he is wrong "because I say so." And that proof was missing.

Perhaps Casimir expresses it best in summing up the argument between Einstein and Bohr [14]:

> Is there really a difference of opinion between Einstein and Bohr? Is it not a question of words? When Bohr says that quantum mechanics offers a complementary description is that not tantamount to saying that from a classical point of view the description is incomplete? And is that not exactly what Einstein is complaining about? Personally, I think it is a legitimate use of language to say that the limited applicability of classical concepts to atomic and subatomic phenomena shows that the quantum mechanical description is incomplete, but that does not mean that there is no serious difference between Bohr and Einstein. Bohr argues that the quantum mechanical description is as complete as it can possibly be and he is ready to accept the limitations of our pictures of reality and of our language, ready also to renounce strict causality such as we find it in classical mechanics and electrodynamics.... New forces may be revealed, surprising new phenomena may be brought to light, but Bohr considered it impossible that they would ever transgress the limitations imposed by quantum mechanics. Einstein on the other hand, though admitting that quantum mechanics is a powerful discipline that provides a valid description of many phenomena, was convinced that one should search for something beyond, for a theory that would to a certain extent re-establish the notions of classical physics.
>
> A simple analogy may be in order. Thermodynamics is a powerful discipline, providing a satisfactory description of many phenomena, but we know now that behind thermodynamics there are innumerable atoms and molecules at work, and physicists have successfully looked for phenomena where the atomic structure reveals itself. Shouldn't we in a similar way look for something behind the statistical laws of quantum mechanics? Bohr's answer would certainly be an emphatic NO!

While this doesn't completely cover the arguments it is as good a summary as one could perhaps really look for. Indeed, the issues of causality and reality were very important to Einstein, and he was not very accepting of Bohr's need to discard these principles.

Consequently, his answer to the Casimir analogy would be an emphatic YES!

Further Considerations

Schrödinger clearly addressed the difference between classical and quantum mechanical systems in terms of what is knowable. Let us pursue this a little further. In the classical world, particles interact with no loss of knowledge. Consider a pair of spherical objects, such as billiard balls, sitting upon the table. By imparting momentum (both linear and angular) to one ball, the two can be made to interact. Our system has a complete specification of the motion of the two balls after the interaction. In fact, if we know exactly the momentum imparted to the first ball, we can predict exactly the motion of the two balls after the interaction, and measuring the position and/or momentum one of the balls will tell us the corresponding values for the other. Indeed, watching the legendary Willy Mosconi[b] play pool was far different than watching a five-year-old trying to sink the balls. The statement "Know exactly" is the rub. It is important to know very precisely what the initial conditions are if we are to determine the state after the interaction. In fact, in many systems, where many balls may be in play, the dynamics are known to become chaotic, and in chaotic systems very slight differences in initial conditions lead to exponentially diverging trajectories. Thus, one can see the importance of knowing completely the initial conditions.

In a larger system, say with N particles interacting through a central force (the interaction mentioned in the previous paragraph is such a force, albeit a short range one), we have $6N$ variables—three momenta and three positions for each particle (one pair each for each axis of the coordinate system). Our description of the system then has this large number of variables. In statistical physics, it is possible to do a summation over $6(N - 1)$ of the coordinates and obtain a probability distribution function for a single "effective"

---

[b]Willy Mosconi was a consultant on the film *The Hustler*, where Jackie Gleason played the fictional Minnesota Fats and Paul Newman play the young hustler.

particle. However, the single-particle distribution function cannot be decoupled from a two-particle correlation function, except by fiat. The summation process is known as the "BBGKY hierarchy", named for the theorists who developed the approach: Bogoliubov [15], Born and Green [16], Kirkwood [17], and Yvon [18]. The two-particle correlation function represents a mean field approximation of the interaction between the one particle and the other $N - 1$ particles. This correlation was known well before the development of the BBGKY formalism. For example, it is dispatched through Boltzmann's *stosszahlansatz*, or "molecular chaos" assumption. The resulting single particle distribution function satisfies the Boltzmann equation. The BBGKY works for either classical or quantum systems, only requiring the full Hamiltonian and the Liouville equation. The important point is that classical correlation exists and is well known.

In the example above, the two balls are correlated immediately after the interaction, just as is the case for the two interacting quantum systems that become entangled. The difference occurs at two levels. First, when we begin to describe the two balls, or two particles, quantum mechanically, Schrödinger has told us that we are no longer able to describe the system with precision. If we decide to measure the position of the two balls, even before the interaction, a precision measurement means that we can no longer say anything about the momenta within the standard version of quantum mechanics. Thus, the institution of a fundamental uncertainty principle means that we have less available information in the quantum description than in the classical description. Once the two particles have interacted, we become further limited quantum mechanically by the fact that we no longer can talk about the separate properties of each ball. Rather, we have available only a single wave function describing the entangled system. So, when compared with the classical system, we now have far less information about the properties of the two particles and can no longer describe the dynamics of the individual particles. In this sense, the description of the quantum entangled particles can be considered to be less than complete. Moreover, the idea that complementarity arises simply from classical physics with a clear connection to quantum physics is no longer tenable.

The situation is more complex than just that expressed above. When we talk about the quantum description of the two particles prior to the interaction, we are still limited. If we express the system in a position space wave function and measure precisely the positions of the two particles, we can say nothing about the momenta. In some sense the momenta are hidden from us so that we cannot gain information about them. Similarly, if we express the wave functions in a momentum representation (Fourier transforms of the position dependent wave functions) and make precise measurements of the two momenta, we can say nothing about the positions. In some sense the positions are hidden from us so that we cannot gain information about them. This situation has led many to ascribe the need for hidden variables in a full description of the quantum system, and this need is often ascribed to EPR. EPR considers the possible ways of writing the wave function and finds two possible results for the reduced wave function of the second particle. Standard quantum mechanics forbids this result for non-commuting operators, so part of the EPR results must be hidden from the observer if one tries to reconcile the two views. So, if one was to find a more complete quantum description, the argument goes that these hidden variables would become knowable by the observer. The fact that they are hidden is further "proof" that the standard quantum mechanics is incomplete. To say that this is arguable is perhaps a strong understatement. We have to remember, that Born and Heisenberg had claimed, at the 1927 conference, that quantum mechanics was closed, and therefore complete. Yet, one has to say that the numerous developments in the field of quantum mechanics subsequent to the 1927 conference must have brought this statement into significant question.

Schrödinger on Measurement

That there are problems with the Copenhagen–Göttingen view of quantum mechanics was illustrated by the EPR problem and by Schrödinger's overview of the present state of quantum mechanics. While we have treated his "cat" paradox as well as the concept of entanglement, there is more. He also extensively laid out some principles which the measurement process must satisfy, but these also raise more questions. As he said [2],

> The rejection of realism has logical consequences. In general, a variable *has* no definite value before I measure it; then measuring does *not* mean ascertaining the value that it has. But then what does it mean?

In the Copenhagen–Göttingen orthodoxy, as we have pointed out several times, reality is assumed to be established only after measuring the system. It is this view that creates some complication in interpretation, and a number of conflicting needs. To approach the question, we need first to define just what a measurement is supposed to be. For this purpose, we can write [2]:

> The systematically arranged interaction of two systems (measured object and measuring instrument) is called a measurement of the first system, if a directly-sensible variable feature of the second (pointer position) is always produced within certain error limits when the process is immediately repeated (on the same object, which in the meantime must not be exposed to any additional influences).

Before this first measurement, it might be supposed that one had some prediction for the result from quantum theory. If one supposes that it is described by a wave function, how could the measurement be repeated? The standard view is that the wave function is collapsed, perhaps by projection into one of the eigenstates of the measuring system [19]. It is then impossible to repeat this measurement on the original system. If we define the entire group of possible results of measurements that can be predicted from the theory, Schrödinger suggested that we call this the "expectation-catalog." Such a quantity is completely different than the wave function itself. Nevertheless, the actual measurement that is made changes this catalog of possible results. If the measurement procedure is known, then we can predict from the theory what results are likely to occur, and the actual measurement at once reduces the expectations to something in the neighborhood of the actual result. This entire process says that [2] any measurement suspends the law that otherwise governs continuous time-dependence of the ψ-function and brings about in it a quite different change, not governed by any law but rather dictated by the result of measurement.

The discontinuity of the expectation-catalog due to measurement is *unavoidable*, for if measurement is to retain any meaning at all than the [expected] *measured value*, from a good measurement, *must* [result].

The discontinuous change [collapse] is certainly not governed by the otherwise valid causal law, since it depends on the measured value, which is not predetermined.

But, now, if we want to repeat the measurement, this cannot be done on the already measured system. Hence, we are forced to create multiple copies of the original system and make identical measurements on each of the copies. This creates some additional factors of importance [2]:

Every measurement is for its system the first. Measurements on separated systems cannot directly influence each other—that would be magic. Neither can it be by chance, if from a thousand experiments it is established that virginal measurements agree.

How can we make copies of the original system? *If the system has no reality until it is measured, then we have insufficient information with which to make copies of this system.* If we cannot make copies, we cannot make additional measurements. Now, we are in a contradictory loop. We need multiple copies with which to make multiple measurements, but the standard Copenhagen–Göttingen orthodoxy precludes us from having sufficient information with which to make multiple copies.

There is another point here. We have admitted to the need to have copies of the original system in order to make multiple measurements, but there is also the need to have the measuring system in precisely the same original state prior to the measurement. Only if the instrument is returned to its precise original state can it be assured that the same excitation from the system to be measured will produce the same measured value. As an alternative, one could consider having multiple copies of the measurement apparatus, all of which must be in the same initial state. Schrödinger asks whether, in fact, this would mean that all of the measurement systems are entangled [2]. As the measurement is made, it is clear that each system to be measured and each measuring system become entangled, and it is this entangled wave function which is projected onto the particular state that results from the measurement. If this is to occur in the same manner for each copy of the original system as well as for each measurement apparatus, it may well be required that the measurement systems must be correlated classically, and hence entangled quantum mechanically. Schrödinger admits that [2]:

> a *prediction* of this extent is thus utterly impossible according to the teaching of Q.M., which we here follow out to its last consequences. . . . ; what answer a system *would have given* to the experimenter . . . has nothing to do with an actual measurement and so, from our epistemological standpoint, does not concern us.

This raises additional questions about what measurement really means, and that brings us back to the problem of the cat. What if the observer cannot communicate the results of the measurement. Schrödinger puts the question in this way [2]:

> And what if we *don't* look? Let's say it was photographically recorded and by bad luck light reaches the film before it was developed. Or we inadvertently put in black paper instead of film. Then indeed have we not only not learned anything new from the miscarried experiment, but we have suffered loss of knowledge.

The latter follows from the fact that the interaction between the system and the measuring apparatus did occur, and the collapse of the wave function into an eigenstate of the measuring apparatus did occur, but the fact that the result was not observable means that no one knows the result. So, knowledge of the result must follow by a second step—the inspection of the results. If the measurement *and* the inspection are to define reality for the quantum system, we might infer that no measurement really occurred for the dead observer who cannot communicate the information to subsequent interested parties. To the casual observer of the situation (not of the experimental results), it might seem that we are in an infinite loop of having to define at what level of cognizance reality occurs, and when do we biologically sense the actual result. To be sure, these arguments have existed for decades, but it seems to thrust the question into the realm of mind-body existence. Perhaps a competent physicist would not want to go there. A different view is expressed well by Penrose [20]:

> To such as myself (and Einstein and Schrödinger too—so I am in good company), it makes no sense to use the term "reality" just for objects that we can perceive, such as (certain types of) measuring devices, denying that the term can apply at some deeper underlying level. Undoubtedly the world is strange and unfamiliar at the quantum level, but it is not "unreal." How, indeed, can real objects be constructed from unreal constituents? Moreover, the mathematical laws that govern the quantum world are remarkably precise—as precise as the more familiar equations that control the behavior of macroscopic objects—despite the fuzzy images that are conjured up by such descriptions as "quantum fluctuations" and "uncertainty principle."

Bohr Capitulates

In the arguments over the EPR ideas, Bohr seemed to hold fast to his view that quantum mechanics was complete, as several authors have noted. Yet, in 1937, he seemed to waver on this hard-line position. Here, he said [21],

> the present formulation of quantum mechanics in spite of its great fruitfulness would yet seem to be no more than a first step in the necessary generalization of the classical mode of description, justified only by the possibility of disregarding in its domain of application the atomic structure of the measuring instruments themselves in the interpretation of the results of experiment. For a correlation of still deeper lying laws of nature involving not only the mutual interaction of the so-called elementary constituents of matter but also the stability of their existence, this last assumption

can no longer be maintained, as we must be prepared for a more comprehensive generalization.

With this remark, he seems to be allowing that perhaps there is a deeper set of underlying physical descriptions that would modify the current view of quantum mechanics. But he still ties it to his precious complementarity, of which he says [21],

> this circumstance . . . forces us to replace the ideal of causality by a more general viewpoint usually termed "complementarity."

Once more, it seems that he is extending his view to cover even more concepts of classical mechanics within the philosophy. Certainly, and particularly, after World War II, many have tried to develop alternative views of quantum mechanics which remove objections raised by Bohr. Beller describes this as [22]:

> What seems impossible with common-sense analysis is, in fact, achieved by mathematical ingenuity. Similarly, Bohr's philosophy of complementarity is permeated by "impossibility" proofs, which are based on an analysis of common-sense ideas (and their "extension"—classical concepts). These "proofs" forbid the very possibility of objective, deterministic, observer-independent ontology. Yet mathematical elaborations of quantum mechanical formalism and interpretation alternative to the Copenhagen one demonstrate that such ontologies are possible, despite Bohr's prohibitions.

Is it not then quite possible that Einstein was correct, and Bohr again was wrong as he so often was?

References

1. E. Schrödinger, *Naturwissen.* **23**, 807, 823, 844 (1935).
2. J. D. Trimmer, *Proc. Am. Philos. Soc.* **124**, 323 (1980).
3. E. Schrödinger, *Proc. Cambridge Philos. Soc.* **31**, 555 (1935); **32**, 446 (1936).

4. C. P. Enz, *No Time to be Brief: A Scientific Biography of Wolfgang Pauli* (Oxford Univ. Press, Oxford, 2002).

5. L. Rosenfeld, in *Neils Bohr: His Life and Work as Seen by his Friends and Colleagues*, ed. by S. Rozenthal (North Holland, Amsterdam, 1967), reprinted in J. A. Wheeler and W. H. Zurek, *Quantum Theory and Measurement* (Princeton Univ. Press, Princeton, 1983).

6. N. Bohr, *The Solvay Meetings and the Development of Quantum Mechanics*, presented at the 12th Solvay Conference, 1961, available from the Neils Bohr Library.

7. A. Einstein, B. Podolsky, and N. Rosen, *Phys. Rev.* **47**, 777 (1935).

8. J. von Neumann, *Mathematische Grundlagen der Quantenmechanik* (Springer, Berlin, 1932).

9. N. Bohr, *Nature* **136**, 65 (1935), reprinted in J. A. Wheeler and W. H. Zurek, *Quantum Theory and Measurement* (Princeton Univ. Press, Princeton, 1983).

10. N. Bohr, *Phys. Rev.* **48**, 696 (1935).

11. J. T. Cushing, *Quantum Mechanics: Historical Contingency and the Copenhagen Hegemony* (Univ. Chicago Press, Chicago, 1994), p. 25.

12. A. Whitaker, *Einstein, Bohr and the Quantum Dilemma* (Cambridge Univ. Press, Cambridge, 1996), p. 168.

13. F. Selleri, in American Institute of Physics Oral Histories: https://www.aip.org/history-programs/niels-bohr-library/oral-histories/28003-1

14. H. B. G. Casimir, in *The Lesson of Quantum Theory: Niels Bohr Centenary Symposium, 1985*, ed. by J. de Boer, E. Dal, and O. Ulfbeck (North-Holland, Amseterdam, 1986), pp.19–20.

15. N. N. Bogoliubov, *J. Phys. Sowj. Union* **10**, 256 (1946).

16. M. Born and H. S. Green, *Proc. R. Soc. London, A* **188**, 10 (1946).

17. J. G. Kirkwood, *J. Chem. Phys.* **14**, 180 (1946).

18. J. Yvon, *Act. Sci. Ind.* **542, 543** (Herman, Paris, 1937).

19. R. Omnès, *The Interpretation of Quantum Mechanics* (Princeton Univ. Press, Princeton, 1994).

20. R. Penrose, *Shadows of the Mind: A Search for the Missing Science of Consciousness* (Oxford Univ. Press, Oxford, 1994), Chapter 6.

21. N. Bohr, *Philos. Sci.* **4**, 289 (1937).

22. M. Beller, *Osiris* **14**, 252 (1999).

Chapter 7

The Rising Storm

After the 1933 Solvay conference, the conferences were suspended, not to be begun again until after the coming global conflagration. Indeed, after EPR in 1935, the lively discussion over the philosophical foundations of quantum mechanics also seemed to disappear for something like a decade. This does not mean that science did not advance, particularly in the area of atomic and nuclear physics. This field exploded with earth shaking effects, but in the postwar world of quantum mechanics, work by David Bohm would create a fire storm.

Work on "splitting" the atom began as early as 1934, when Enrico Fermi began to bombard uranium atoms with neutrons. From then until a few years later, it was thought that atoms whose atomic numbers were greater than uranium (92) were produced by this process [1]. Not everyone agreed with Fermi, and the German chemist Ida Noddack suggested that "it is conceivable that the nucleus breaks up into several large fragments, which would of course be isotopes of known elements but would not be neighbors of the irradiated element" [2]. Later, in 1938, Otto Hahn, a German chemist and his assistant bombarded uranium with neutrons and discovered a light element [3], which they later identified as barium. Earlier, Hahn had assisted the Jewish chemist, Lisa Meitner, to escape

The Copenhagen Conspiracy
David Ferry
Copyright © 2019 Pan Stanford Publishing Pte. Ltd.
ISBN 978-981-4774-75-8 (Hardcover), 978-1-351-20723-2 (eBook)
www.panstanford.com

to the Netherlands and on to Sweden. Before publishing his new results, he traveled to Copenhagen to discuss the results with Bohr and Meitner. Meitner and her nephew, Otto Frisch, then worked out the theory to confirm these results [4]. This work on neutron bombardment had been undertaken, as Niels Bohr had suggested that spontaneous uranium fission would be a negligible process. However, the Soviet physicists Flerov and Petrzhak demonstrated that spontaneous fission of uranium could occur [5]. As news of the experiments spread across the Atlantic, another refugee, Leó Szilárd, realized that the number of neutrons that could be emitted would allow for a chain reaction, in which one fission process would trigger additional decay processes. Together with Edward Teller and Eugene Wigner, also Hungarian transplants, he contacted Einstein and urged him to take the information to the American president Roosevelt. This eventually led to the American Manhattan Project. By 1942, the Manhattan Project had major facilities at Oak Ridge (Tennessee), Los Alamos (New Mexico), and Hanford (Washington), as well as 16 other smaller facilities. The subsequent history is well known. As a by-product, the fact that a very large number of US and British nuclear scientists had quit publishing, alerted their Russian counterparts that something was going on, which the Russian government confirmed by a number of methods. This led to their own project shortly after the German invasion of Russia.

While the Manhattan Project was led by the military, the number of civilian scientists involved was remarkable. But other scientists contributed to other tasks, with almost 10,000 being pulled into the war effort. Major laboratories were created at several universities, under civilian control of the Office of Scientific Research and Development. Perhaps the most famous was the Radiation Laboratory at MIT, which was concerned with electromagnetic generation and propagation. This massive military-industrial complex would have remarkable effects on science, as it pushed the leading scientists into the public eye and made them instantly recognizable. This was a property that previously had only been given to Einstein.

The Positivist Revival

The Solvay physics conferences resumed in 1948, and this version was held in Brussels, Belgium. The topic was elementary particles, but there would be an undercurrent of quantum mechanics. Of interest is that the attendees now included the new generation of physics, men and women who had led the scientific programs during the war. While old timers like Schrödinger, Bohr, Pauli, Dirac, and Kramers were there, most attendees could be considered to be from a new generation. These included J. Robert Oppenheimer, Edward Teller, Lisa Meitner, Felix Bloch, Rudolf Peierls, Hendrik Casimir, Leon Rosenfeld, and Leon van Hove. Other practitioners of quantum mechanics who were not there, still have leaped into the public consciousness. These included Norbert Wiener, Eugene Wigner, John Wheeler, Walter Elsasser, and Maria Goeppert-Mayer. It is not important whether or not they attended the Solvay conference; what is important is that this group would lead the community of people interested in quantum mechanics, and it would cement the acceptance of Bohr's positivist philosophy. Franco Selleri [6] later asked why this was

> in that way I discovered how strong was the opposition to the Copenhagen approach, because I soon found out that Einstein was against, Planck was against, Schrödinger was against, de Broglie was against, Ehrenfest was against. It was not just one man, it was a lot of very important people. And then the problem, which is still open today ... is: how is it possible that the existing scientific community was born from such a splitting in the group of creators and founders? You can say basically that twelve of the most important people were split 50:50 against and in favor of the Copenhagen quantum mechanics. Now from a situation like this normally in any human activity you would find the later cultural community also split, but not in physics. In physics you have 99:1. Fifty-fifty for the founding fathers, but 99 to 1 for the people who are active in research [today]. How was it possible, through repression, and control of positions and publications? And then also a lot of dogma exists. People do not dare opposing [*sic*] important ideas. They feel that it is too dangerous, but they do not dare anyway.

One can suspect there are probably a great many contributors to the answer that could be given to Selleri, but there is one very important answer that contributed significantly—students. Einstein produced no students. Schrödinger produced no students of great significance. De Broglie only produced one significant student—Bernard d'Espagnat—after the war. On the other hand, J. Robert Oppenheimer, although an American, did his doctoral work at Göttingen under Born beginning in 1926, and so was present at the founding of quantum mechanics. Edward Teller, a Hungarian, received his doctorate at Leipzig under the direction of Heisenberg. Norbert Wiener, an American, studied at Göttingen in 1926 as a Guggenheim scholar. Eugene Wigner, a Hungarian, pursued his undergraduate work in Berlin and studied quantum mechanics with Schrödinger, but studied in Göttingen and eventually became known as a positivist. Felix Bloch, born in Switzerland, received his degree in 1927 from the ETH, then studied with Heisenberg in Leipzig, Pauli in Zürich, Bohr in Copenhagen, and Fermi in Rome. Rudolf Peierls studied in Zürich with Heisenberg and Pauli. Hendrik Casimir received his doctorate under Ehrenfest, but then studied in Copenhagen with Bohr and Zürich with Pauli. Léon Rosenfeld received his doctorate in Liége and then went to Copenhagen to study with Bohr. John Wheeler was a collaborator of Bohr, especially when the latter visited America. Léon van Hove was of a new generation, receiving his doctorate in Brussels after the war, but then going to IAS to work with Oppenheimer. Walter Elsasser and Maria Goeppert-Mayer were both students of Born in Göttingen and overlapped with Oppenheimer during that period. Thus, it seems that the second generation of scientists working in quantum mechanics all had strong input from the Copenhagen–Göttingen crowd. It is not surprising then that the new leadership in physics was a fully indoctrinated set of acolytes to this philosophy.

As may be expected, this sudden assumption of leadership by the positivists came as a shock to Schrödinger. In a letter to Einstein, he remarked [7],

> It seems to me that the concept of probability is terribly mishandled these days. Probability surely has as its substance a statement as to whether something *is* or *is not* the case—an

uncertain statement, to be sure. But nevertheless it has meaning only if one is indeed convinced that the something in question quite definitely *is* or *is not* the case. A probabilistic assertion presupposes the full reality of its subject.... But the quantum mechanics people sometimes act as if probabilistic statements were to be applied *just* to events whose reality is vague.

The conception of a world that really exists is based on there being a far-reaching common experience of many individuals, in fact of all individuals who come into the same or a similar situation with respect to the object concerned. Perhaps instead of "common experience" one should say "experiences that can be transformed into each other in a simple way." This proper basis of reality is set aside as trivial by the positivists when they always want to speak only in the form: if "I" make a measurement then "I" "find" this or that. (And that is the only reality.)

For it is just because they prohibit our asking what really "is," that is, which state of affairs really occurs in the individual case, that the positivists succeed in making us settle for a kind of collective description. They accuse us of metaphysical heresy if we want to adhere to this "reality." That should be countered by saying that the metaphysical significance of this reality does not matter to us at all.

It is to this letter that Einstein replied [8]:

You are the only contemporary physicist, besides Laue, who sees that one cannot get around the assumption of reality—if only one is honest. Most of them simply do not see what sort of risky game they are playing with reality—reality as something independent of what is experimentally established. They somehow believe that the quantum theory provides a description of reality, and even a *complete* description.... Nobody really doubts that the presence or absence of the cat is something independent of the act of observation. But then the description by means of the ψ-function is certainly incomplete, and there must be a more complete description. If one wants to consider the quantum theory as final (in principle), then one must believe that a more complete description would be useless because there would be no laws for it. If that were so then physics could only claim the interest of shopkeepers and engineers; the whole thing would be a wretched bungle.

Earlier, Einstein attempted to explain more thoroughly his objections to the completeness of quantum mechanics. Here, he wrote [9],

> I consider a free particle described at a certain time by a spatially restricted ψ-function (completely described—in the sense of quantum mechanics). According to this, the particle possesses neither a sharply defined momentum nor a sharply defined position. In which sense shall I imagine that this representation describes a real, individual state of affairs? Two possible points of view seem to me possible and obvious....
>
> (a) The (free) particle has a definite position and a definite momentum, even if they cannot both be ascertained by measurement in the same individual case. According to this point of view, the ψ-function represents an incomplete description of the real state of affairs.
>
> This point of view is not the one physicists accept. Its acceptance would lead to an attempt to obtain a complete description of the real state of affairs as well as the incomplete one, and to discover the physical laws for such a description. The theoretical framework of quantum mechanics would then be exploded.
>
> (b) In reality the particle has neither a definite momentum nor a definite position; the description by ψ-function is in principle a complete description. The sharply-defined position of the particle, obtained by measuring the position, cannot be interpreted as the position of the particle prior to the measurement.... This is presumably the interpretation preferred by physicists at present; and one has to admit that it alone does justice in a natural way to the empirical state of affairs expressed in Heisenberg's principle....
>
> According to this point of view, two ψ-functions which differ in more than trivialities always describe two different real states of affairs, even if they could lead to identical results when a complete measurement is made.
>
> If one asks what, irrespective of quantum mechanics, is characteristic of the world of ideas of physics, one is first of all struck by the following: the concepts of physics relate to a real outside world, that is, ideas are established relating to things such as bodies, fields, etc., which claim a "real existence" that is independent of the perceiving subject....

> I now make the assertion that the interpretation of quantum mechanics is not consistent with [the above principle of physics].

Einstein then goes through an argument quite similar to that in EPR, which demonstrates that it is difficult to align alternative (b) with the reality that is commonly accepted in the rest of physics. This leads him to prefer alternative (a) which means that the ψ-function is an incomplete description of reality, and it must be expected that quantum mechanics will be replaced at some later date with a more complete theory.

From the above, it appears that Einstein and Schrödinger are standing alone in their philosophical view of quantum mechanics (we recall that de Broglie defected to the Göttingen interpretation following the 1927 conference). The Copenhagen–Göttingen view of the world was winning the contest, but time would bring several storms into their way, and these would challenge once again the accepted interpretation.

David Bohm

David Bohm was an American born of immigrant parents in Pennsylvania in 1917. He pursued his undergraduate degree at Penn State, and then moved to Caltech in 1939. Here, he was disappointed in the environment, saying [10],

> there I was rather unhappy because the atmosphere was very oppressive. They constantly were giving exams and competing and were not interested in what the subject was about.

So, in 1941, he moved to Berkeley and began working with J. Robert Oppenheimer. Here, Bohm worked on the physics of ionized gases, termed *plasmas*. When the war broke out, and Oppenheimer went to the Manhattan project, Bohm remained at Berkeley to complete his degree. He finished his thesis in 1943, working on scattering calculations of protons and neutrons in the plasma. His thesis was immediately classified and he could no longer read it, or even his own notes, since he did not have a security clearance. Oppenheimer had to intercede with the university to assure the administrators that he had satisfied the degree requirements. Bohm then moved to Oak Ridge, where he did achieve the necessary clearances to contribute to important nuclear research. After the war, he became an assistant professor at Princeton in 1947.

At Princeton, Bohm became part of a sizable group in theoretical physics, but he wasn't overall happy either with the town or with the atmosphere [10]:

> the environment was flat and uninteresting ... the town was a bit staid, you know; it wasn't very interesting ...
>
> Wigner and I were not hostile, but we're not really close to each other in our way of thinking. That is, Wigner is far more formal. And Wheeler I could talk to a bit you see. I found the general atmosphere a little bit difficult in the same way as Cal Tech. There's a lot of status consciousness ... and this interfered with freedom of thought.

Nevertheless, he had several good students. Although Gross worked with him, he finished under another advisor, but David Pines did his doctorate with Bohm.

In the postwar time, there was a political concern with communism and communists infiltrating the US science, cultural, and political establishment. This was brought to a head by what was called the House Un-American Activities Committee. While it was chaired by Edward Hart, Senator Joseph McCarthy was a regular attendee and became famous for directly asking people who were testifying if they were, or had been, a communist. In 1948, they went after Hollywood and this led to the famous "black list." In 1949, it seemed that they were after Oppenheimer, and called on Bohm and other former colleagues to testify before the committee. Bohm invoked his fifth amendment right to not testify and refused to attend the hearings. He was arrested for contempt of congress in 1950, but was acquitted the following year. Nevertheless, Princeton decided not to give tenure to Bohm, and he was allowed to serve out the last year of his contract, providing he did not come to campus. Perhaps they were just as glad to be rid of him, as he was Jewish. This was a problem at some of the elite universities. For example, Richard

Feynman was not admitted to Columbia, because they had a quota on Jewish students, and he was almost not admitted to graduate work at Princeton because of this fact.

As a result of his charge from Congress and his dismissal from Princeton, Bohm became a persona not grata at other universities. In October 1951, he gathered his belongings and left for Brazil. Upon arrival, the U. S. Consulate [11]

> took my passport and said that I wouldn't get it back except for return to the United States. Fortunately, I had an identity card so I could stay there.

It was in Brazil, that he worked on his contribution to quantum mechanics—the causal theory. While at Princeton, he had written a book on quantum mechanics in which he expressed the accepted Bohr interpretation. This left him unsatisfied, and he felt that there was something else there, that needed to be said, and this culminated in the causal theory, which we return to in a subsequent section. In Brazil, he worked and taught at the Universidade de São Paulo, and he became an attraction as many scientists traveled to Brazil to talk with him. In 1955, his Jewishness facilitated a move to Israel, where he met his wife. Here, he worked with Yakir Aharonov, a student at the Technion in Haifa. After two years, he moved again, this time to the United Kingdom. This time, Aharonov moved with him and completed his degree with Bohm in Bristol. Finally, in 1961, Bohm became a professor at Birkbeck College, part of the University of London. He would remain there for the rest of his life.

The "Usual" Understanding

Before proceeding to the causal theory, and with it, the return to the world of de Broglie, it is useful to describe what the so-called *"usual" theory* entailed. This Copenhagen–Göttingen interpretation can best be said to be [12]

> the physical state of an individual system is completely specified by a wave function that determines only the probabilities of actual

results that can be obtained in a statistical ensemble of similar experiments.

Quantum-mechanical probabilities are regarded ... as only a practical necessity and not as a manifestation of an inherent lack of complete determination in the properties at the quantum level.

... Bohr and others have suggested an additional assumption, namely, that the process of transfer of a single quantum from the observed system to the measuring apparatus is inherently unpredictable, uncontrollable, and not subject to a detailed rational analysis or description.

Wigner would later put it somewhat differently [13]:

There are two epistemological attitudes toward this [the state vector]. The first attitude considers the state vector to represent reality, the second attitude regards it to be a mathematical tool to be used to calculate the probabilities for the various possible outcomes of observations. It is not easy to give an operational meaning to the difference of opinion which is involved because, fundamentally, the realities of objects and concepts are ill defined. One can adopt the compromise attitude according to which there is a reality to objects but quantum mechanics is not concerned therewith. It only furnishes the probabilities for the various possible outcomes of observations or measurements—in quantum mechanics these two words are used synonymously.

Wigner admits the viewpoint of Einstein, but draws away from it in the end. These two viewpoints illustrate that there was no commonly accepted version of the standard approach to quantum mechanics, and hence it is very difficult to pin anyone down on just what the "Copenhagen Interpretation" means. Bohr was, of course, incapable of providing a simple, easily understandable answer, even if he had one.

The strict view expressed by Bohm fits with the description provided by Einstein earlier in this chapter. It is this description that Einstein says does not fit with the standard interpretation of how physicists characteristically approach the world. It is this view that is unacceptable to him. Wigner is a little more wishy-washy about the presence of a reality prior to the measurement and allows as how such a reality might be important, but he lapses at the end into the "compromise" which leads him to the same final view expressed by Bohm. Others would go further and state [14],

> Quantum mechanics does not describe a system in itself, but only deals with the results of actual observations on it. Due to the central role of observation, an understanding of the process of measurement is of a fundamental importance in quantum mechanics and it is intrinsically connected with the interpretation of the theory itself.

This is almost a quote from Heisenberg's first paper and most clearly expresses the view that has become expected from Copenhagen. It is Mach's prescription almost exactly. It is clear that if one follows this prescription, then questions about trying to understand what is exactly going on with the variables and particles in the physical system under investigation can be said to be irrelevant. That is, physics is mostly irrelevant, only the perception of an event is meaningful. This brings us back to the problems raised with Schrödinger's cat. If the observer dies before communicating the result, has a measurement really been made? And this leads to the tragedy described by Schrödinger discussed in the last chapter— not only have we learned nothing from the attempt at observation, we have also suffered a loss of information about the system due to the collapse of the wave function. One is only left scratching one's head over what this means for the entire world of scientific understanding.

The Causal Theory

David Bohm put together his causal theory after being unhappy with the general state of quantum mechanics as he found it during the time he was writing his book [15]. He termed it the causal interpretation. Then, he discovered the de Broglie view of quantum mechanics [16], as well as the hydrodynamic approach followed by Madelung [17], and became aware that de Broglie had backed away from his own theory under questioning by Pauli at the 1927 Solvay conference. In Bohm's view [12],

> all of the objections of de Broglie and Pauli could have been met if only de Broglie had carried his ideas to their logical conclusion.

Bohm goes on to criticize the usual interpretation of quantum theory on his fundamental approach [12]:

> it requires us to give up the possibility of even conceiving precisely what might determine the behavior of an individual system at the quantum level, without providing adequate proof that such a renunciation is necessary. The usual interpretation is admittedly consistent; but the mere demonstration of such consistency does not exclude the possibility of other equally consistent interpretations.

For Bohm, another such consistent approach is his causal theory, and he admits that such an alternative approach could possess parameters and variables beyond those of the usual theory. He refers to these additional parameters as "hidden variables." To be sure, hidden variables were discussed already by Einstein and Schrödinger. These occur when one projects from classical mechanics with its easily describable sets of conjugate variables, such as position and momentum. In quantum mechanics, where these two variables are not capable of being simultaneously measured, we already have less information than in classical mechanics. Bohm points out that such hidden variables are an intrinsic part of classical statistical physics with its ensembles of systems. Because atoms are not individually observable in Brownian motion, it is natural to think of them as described by hidden variables. He then connects with the incompleteness described by Einstein with the following comments [12]:

> If there are hidden variables underlying the present quantum theory, it is quite likely that in the atomic domain, they will lead to effects that can be described adequately in the terms of the usual quantum-mechanical concepts; while at a domain associated

with much smaller dimensions ... the hidden variables may lead to completely new effects not consistent with the extrapolation of the present quantum theory down to this level.

More importantly, the statements of the usual interpretation of quantum mechanics are not extensive enough to imply a unique mathematical formulation. The Hamiltonian operator can be modified in almost any manner without violating the interpretation. As Bohm puts it [12],

as long as we accept the usual interpretation of quantum theory, we cannot be led by any conceivable experiment to give up this interpretation, even if it should happen to be wrong. The usual physical interpretation therefore presents us with a considerable danger of falling into a trap, consisting of a self-closing chain of circular hypotheses, which are in principle unverifiable if true.

Bohm developed the causal theory in full compliance with the Schrödinger approach, but also with the de Broglie approach. It is assumed that there exists a wave function for a single system, which may consist of a single particle. So, in addition to the wave function, there is a particle. The squared magnitude of the wave function represents the probability of finding the particle at any value of **x** (in three dimensions). Since we cannot determine the actual value of the position of the particle, although it has a unique position at each instant of time, we have to admit that this position is *hidden* from the observer. Hence, the position of the individual particle is a hidden variable. Correspondingly, the particle has a momentum which will also be inaccessible to the observer. So, the momentum is also a hidden variable. Bohm describes his wave function in the same manner as Madelung [17]. That is, we may write the wave function as

$$\psi(\mathbf{x}) = A(\mathbf{x}) \exp\left[i\frac{S(\mathbf{x})}{\hbar}\right], \qquad (7.1)$$

where the amplitude A and the reduced phase S are real functions. Obviously, the probability is given by A^2 and nothing that appears in the phase can be seen from the probability. The function S is a solution of the normal Hamilton–Jacobi equation and can be recognized as being connected to the classical ideas of the *action* (integral over time of the energy). If one now uses the wave function 7.1 in the Schrödinger equation, one arrives at a complex equation, in which the real and imaginary parts must separately balance. This gives two equations

$$\frac{\partial A}{\partial t} = -\frac{1}{2m} \left[A \nabla^2 S + 2(\nabla A) \cdot (\nabla S) \right], \tag{7.2}$$

$$\frac{\partial S}{\partial t} = -\left[\frac{(\nabla S)^2}{2m} + V(\mathbf{x}) - \frac{\hbar^2}{2m} \frac{\nabla^2 A}{A} \right]. \tag{7.3}$$

If we write the probability as $P = A^2$, then the first of these equations can be rewritten as

$$\frac{\partial P}{\partial t} + \nabla \cdot \left(P \frac{\nabla S}{m} \right) = 0. \tag{7.2a}$$

This last equation (7.2a) can be recognized as the continuity equation for probability. The term in parentheses must be the probability current for this identification. This also allows us to recognize that the second term in these parentheses is the probability velocity $\nabla S/m = \mathbf{v}(\mathbf{x})$. Hence, we can also recognize the momentum. With these relations, it is clear that the continuity equation is consistent with the view that the probability density P is the probability density for the particle(s).

If we let Planck's reduced constant $\hbar \rightarrow 0$, then Eq. 7.3 also has a clear meaning. The left-hand-side term is just the energy so that the first two terms on the right are the kinetic energy and the potential energy. The identification of the derivative of the spatial variation of S with the momentum is consistent with this view. The first term on the right-hand side of Eq. 7.3 is the particle kinetic energy. Then what is the third term on the right-hand side of Eq. 7.3? In the quantum case, it appears that the particle moves under a force which is not simply the classical force deriving from the potential energy. The third term is an additional potential that is purely quantum mechanical in nature, which is precisely the description that Kennard had given earlier [18]. Bohm refers to this term as the "quantum mechanical" potential [12]:

$$U(\mathbf{x}) = -\frac{\hbar^2}{2m}\frac{\nabla^2 A}{A} = -\frac{\hbar^2}{4m}\left[\frac{\nabla^2 P}{P} - \frac{1}{2}\frac{(\nabla P)^2}{P^2}\right]. \tag{7.4}$$

Bohm goes on to say [12],

> Thus, it would seem that we have here the nucleus of an alternative interpretation for Schrödinger's equation.
> Since the force on a particle now depends on a function of the absolute value, $A(\mathbf{x})$, of the wave function, $\psi(\mathbf{x})$, evaluated at the actual position of the particle, we have effectively been led to regard the wave function of an individual electron as a mathematical representation of an objectively real field. This field exerts a force on the particle in a way that is analogous to, but not identical with, the way in which an electromagnetic field exerts a force on a charge.... In the last analysis, there is, of course, no reason why a particle should not be acted upon by a ψ-field, as well as by an electromagnetic field, a gravitational field, a set of meson fields, and perhaps by still other fields that have not yet been discovered.

The analogy between the wave function field and the electromagnetic field can be carried through. Just as the vector electromagnetic fields satisfy Maxwell's equations, the scalar wave function field satisfies Schrödinger's equation. In both of these cases, solution of the defining equations allows one to compute the forces that act upon a particle. But it is important to note here that the wave function is now considered to be a field, and discussion of the ψ field is, in fact, a discussion of the wave *function*. If one knows the initial values of the particle's momentum and position, it is then possible, using the time-dependent solutions of these equations, to compute the entire trajectory of the particle, and such approaches have been effective [19]. While the Hamilton–Jacobi equations were mentioned above, we are reminded that simply solving Newton's equations of motion should suffice. In this case, the acceleration equation becomes

$$m\frac{\partial^2 \mathbf{x}}{\partial t^2} = -\nabla\left[V(\mathbf{x}) - \frac{\hbar^2}{2m}\frac{\nabla^2 A}{A}\right]. \tag{7.5}$$

This leads to [12]:

> It is in this connection with the boundary conditions appearing in the equations of motion that we find the only fundamental difference between the ψ-field and other fields.... In order to obtain results that are equivalent to those of the usual interpretation of the quantum theory, we are required to restrict the value of the initial particle momentum to $\mathbf{p} = \nabla S(\mathbf{x})$. From the application of Hamilton–Jacobi theory ... it follows that this restriction is consistent, in the sense that if it holds initially, it will hold for all time.
>
> This probability density is numerically equal to the probability density of particles obtained in the usual interpretation. In the usual interpretation, however, the need for a probability description is regarded as inherent in the very structure of matter ... whereas in our interpretation, it arises ... because from one measurement to the next, we cannot in practice predict or control the precise location of the particle....
>
> The use of statistics is, however, not inherent in the conceptual structure, but merely a consequence of our ignorance of the precise initial conditions of the particle.

The causal approach is not limited to the case of a single particle. In fact, Bohm develops the multi-particle case in detail. Mainly, the wave function is now a function of the positions of each of the particles and is a proper many-body wave function, as described in previous chapters. The important result of this is that the quantum potential is now a function of the positions of all the particles through the many-body wave function. As a consequence, the actual quantum force seen by any single particle depends upon the positions of all the other particles, which gives a nonlinear many-particle effect. If the wave function has the proper antisymmetry under exchange of two particles, then this will prevent any two particles from occupying the identical positions, and the exclusion principle is properly taken into account. But Bohm also notes that [12]

> the uncertainty principle is regarded, not as an inherent limitation on the precision with which we can correctly conceive of the simultaneous definition of momentum and position, but rather as

a practical limitation on the precision with which these quantities can simultaneously be measured.

The Two-Slit Experiment

The two-slit experiment has been one of the most confusing problems in quantum mechanics since the earliest days. As we saw in Chapter 1, Taylor had investigated diffraction effects with very weak light as early as 1909 [20]. Even when the light was sufficiently weak that only a single photon at a time was in the experiment, the diffraction corresponded to the wave picture and the arrival at the photographic plate corresponded to the particle picture. The same had been demonstrated with electrons in an electron microscope under the conditions in which only a single electron was in the machine at one time [21]. As these latter authors state, "A series of experiments was carried out for different electron intensities. . . . The contrast of the fringes remains the same within experimental error." Whether for photons or for electrons, the actual fringe that is built up by the particles arriving at the detector clearly shows the effects of their wave nature as they passed through the diffraction region.

Bohr tried to explain this dual nature of the particle through his complementarity principle, but it is not an either/or situation. It is an "and" situation. One needs the wave *and* the particle, as de Broglie suggested. The actual experiment creates a problem for the usual interpretation of quantum mechanics. Einstein had illustrated the problem with his challenges at the 1927 Solvay conference. The experiment is relatively simple, so that one can pose it clearly. A single particle arrives at a screen with two slits in it. We take the normal direction to the screen to be the z direction, so that the particle can be said to have a momentum $p_z = \hbar k_0$. After passing through one of the two slits it arrives at a second screen, where its position is recorded, for example by a photographic plate in the experiment of Taylor. In the usual interpretation, we describe the particle by a wave that arrives at the screen as $\psi \sim \exp(ik_0 z)$. When this wave passes through the two slits, it undergoes both interference and diffraction and develops a complicated spatial variation that illustrates the interference at the second screen. At this second screen, the wave has a probability in the lateral direction as, e.g., $P(x) = |\psi(x)|^2$. If the experiment is repeated many times, the interference fringe is gradually built up on the photographic plate.

The usual interpretation of quantum mechanics takes little accord of the particle and the origin of the interference becomes difficult to understand. As Bohm puts it [12],

> there may be certain points where the wave function is zero when both slits are open, but not zero when only one slit is open. How can the opening of a second slit prevent the electron from reaching certain points that it could reach if this slit were closed? If the electron acted completely like a classical particle, this phenomenon could not be explained at all. Clearly, then the wave aspects of the electron must have something to do with the production of the interference pattern. Yet, the electron cannot be identical with its associated wave, because the latter spreads out over a wide region. On the other hand, when the electron's position is measured, it always appears at the detector as if it were a localized particle.
>
> The usual interpretation of the quantum theory not only makes no attempt to provide a single precisely defined model for the production of the phenomena described above, but it asserts that no such model is even conceivable. Instead ... it provides ... a pair of complementary models, *viz.*, particle and wave.... While the electron goes through the slit system, its position is said to be inherently ambiguous, so that ... it is meaningless to ask through which slit an individual electron actually passed.

The causal interpretation that Bohm puts forward clarifies this picture considerably. *The same wave function used is that in the usual interpretation, but in the causal picture, this wave function is regarded as a real field.* That is, it is a mathematical representation of a field that provides force upon the particle. The wave function generates this force in agreement with Eq. 7.5. It is the quantum force which leads to the fact that the particles illustrate the interference fringes when they arrive at the detector. If it happens that there are places where the wave function is zero, this leads to an infinitely large force which prevents the particles from approaching these places.

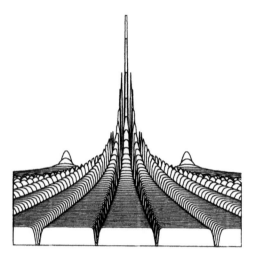

Figure 7.1 The quantum potential for the two-slit experiment, as seen from the screen upon which the interference is found. The two slits are located between the three peaks at the rear. The peaks are regions where the wave function is small and the particles are not allowed to appear. Reproduced from Philippidis et al. with permission of Springer.

Moreover, if one of the slits is closed, the diffraction of the wave function changes. This, in turn, dramatically changes the quantum potential, so that the particle is now allowed to reach places that it had no access to when both slits were open. Hence, the opening or closing of a slit affects the particle indirectly, and only through the changes produced in the quantum potential by such an action. Thus, it is the actual ψ field resulting from the detailed state of the two slits that generates the forces that cause the particle motion corresponding to that particular state. Here, there is no strange mystery about the quantum behavior of such a system.

The role of the quantum potential was investigated by Philippidis, Dewdney, and Hiley [22], where they calculated the quantum potential for the two slit experiment. In Fig. 7.1, we show the quantum potential which they determined from the wave function in the region between the two slits and the screen upon which the interference was developed. These authors noted that the quantum potential appeared to have nonlocal features "which seem to be essential for a proper description of some quantum effects, and it

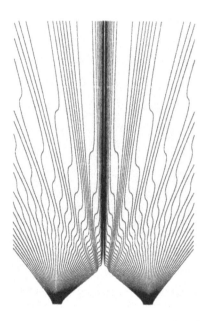

Figure 7.2 Trajectories of particles that are initiated within the two slits (at the bottom of the figure). It can be seen how the quantum potential focuses these particles in a manner to create the interference fringes on the screen (top). Reproduced from Philippidis et al. with permission of Springer.

has no well-defined source." From the equations described above, they found two well-defined wave packets after the two slits, and from these they could numerically find the quantum potential. There are three clear peaks in the quantum potential, with the major peak being located between the two slits. The slits themselves lie between these three peaks. The peaks, as noted by Bohm, are regions where the wave function is small and the particles cannot effectively reach. Clearly, there are troughs in the potential, and the idea is that the particles will be focused into these troughs by the quantum potential. In Fig. 7.2, the corresponding particle trajectories are plotted. The trajectories correspond to different initial conditions within the two slits. One can clearly see how the particles are focused into beams which create the interference fringes on the screen. Clearly, if a particle is excited out of one of the troughs, it moves over the intervening potential into another trough. There are few particles where the potential is high, and a great many particles where the potential is low in the troughs.

EPR Again

Bohm then turned to a consideration of the EPR problem that had been suggested years earlier, and which was discussed in the previous chapter. Here, Einstein, Podolsky, and Rosen had posited a pair of interacting particles which had then separated [23]. If they then measured the momentum of one of the particles, they could infer the momentum of the other, but they then lost any ability to measure the positions of the particles according to the "usual" interpretation. EPR pointed out that they could measure the position of the second particle, and thus asserted that this usual quantum theory is inadequate to give a complete description of the particles or of how the correlation between them is propagated. Bohm suggests that, in fact, one could easily describe the experiment "in terms of a combination of six-dimensional wave field and a precisely definable trajectory in a six-dimensional space" [12]. He then continues,

> if we measure the position of the first particle, we introduce uncontrollable fluctuations in the wave function for the entire system, which, through the "quantum-mechanical" forces, bring about corresponding uncontrollable fluctuations in the momentum of each particle. Similarly, if we measure the momentum of the first particle, uncontrollable fluctuations in the wave function for the system bring about, through the "quantum-mechanical" forces, corresponding uncontrollable changes in the position of each particle. Thus, the "quantum-mechanical" forces may be said to transmit uncontrollable disturbances instantaneously from one particle to another through the medium of the ψ-field.

Now, this instantaneous transmission of the disturbance can lead to conceptual problems. Bohm recognizes this incongruity in the idea, but points out that [12]

> it leads to precisely the same predictions for all physical processes as are obtained from the usual interpretation (which is known to

be consistent with relativity). The reason why no contradictions with relativity arise in our interpretation despite the instantaneous transmission of momentum between the particles is that no signal can be carried in this way. For such a transmission of momentum could constitute a signal only if there were some practical means of determining precisely what the second particle would have done if the first particle had not been observed; and as we have seen, this information cannot be obtained as long as the present form of the quantum theory is valid.

We thus see two important points here. The physical act of measuring the position or the momentum of the first particle introduces the fluctuations into the system. It is this measurement which disturbs the system, and breaks our ability to follow the particles through a causal manner. Secondly, the idea of instantaneous transmission of the momentum fluctuation can be challenged as being questionable only by simultaneously calling the usual interpretation into question. So long as one continues to assert that the usual interpretation is correct, the causal form suggested by Bohm cannot lead to inconsistencies with relativity. This is because the causal form predicts precisely the same results as the usual interpretation. While Einstein allegedly called this "spooky action at a distance," the two forms of quantum theory both suffer from this. In general, it is more common to refer to this as a property of a nonlocal theory. By nonlocality, we generally imply that the causative effect is at one position, while the resulting behavior is at a second position, and this leads to action at a distance. Normally, we suppose that relativity theory rules out any such action at a distance, by limiting the speed at which information can travel to the speed of light.

While EPR had used two simple particles and let them interact, Bohm [15] suggested another method of phrasing the paradox. He suggested that we consider a diatomic molecule, which is split apart by a method that does not affect the spin of each of the two atoms. In the molecule, the total spin is zero, as the two atoms have equal and opposite spin. In the splitting process, this does not change. Once the two atoms have moved apart, there is no further interaction. However, each molecule has both a position and a momentum. Yet, Bohm suggests that the wave function for the two atoms can be written as

$$\psi = \frac{1}{\sqrt{2}} [\psi_+(1)\psi_-(2) - \psi_-(1)\psi_+(2)], \qquad (7.6)$$

where the subscripts refer to either up- (positive) or down- (negative) spin, and the arguments refer to atom 1 or atom 2. But this is an incomplete wave function, and only describes the spin states of the two atoms. One could say that this is a minimal information wave function, but it has in recent years become the de facto wave function for the EPR problem. For example, we know that each atom has a position and a momentum, but this wave function *can tell us nothing about these quantities* even if a measurement is made on the system seeking the position or the momentum. Indeed, the positions and the momenta of the two atoms must be considered as hidden variables, as no experiment on the system as described by Eq. 7.6 can yield any information about the probability distribution of these quantities. Another issue is that a measurement of the spin of either atom can yield only information about a single measurement. It gives either up or down! If we are to believe that the proper wave function gives us the probability distribution of possible values, then Eq. 7.6 is a failure. If we are to adopt the usual interpretation of quantum mechanics, for example, as given by Wigner [13] as "[the wave function] only furnishes the probabilities for the various possible outcomes of observations or measurements" or as given by Prosperi [14] as "Quantum mechanics does not describe a system in itself," then Eq. 7.6 does not represent a valid expression for this system. At best, it is a statement of our lack of specific knowledge about the spin of the two atoms, but lack of knowledge is not a basis upon which to construct a viable model to be used in evaluation of the quantum properties of a system.

Bohm and Aharonov raised further questions about this version of the EPR paradox [24]. The problem with spin is that a measurement of the spin determines only one axis, for example the x-directed part of the spin. While this, for sure, could be designated as the up spin axis, it is not complete. While the measurement gives us a value for the x component of the spin, it cannot give us information about the other two components. In the situation first described by EPR, the complementary pairs of continuous variables, either the position or the momentum, lead to the uncertainty principle for measurements of either quantity, and this holds for both of the particles. In the present formulation, where a measurement may be made on the spin of particle 1, one must address the question of just what physics comes into play such that particle 2 has exactly the opposite spin of which ever direction is selected for the measurement. This is especially a problem when you realize that the other two components of the spin of particle 1 are fluctuating, and yet there is no uncertainty in the result for either particle. And this fluctuation cannot be explained away as a result of disturbances in the measuring apparatus. The conundrum of the Bohm version of EPR is that it has been widely developed as the easier approach to take to describe the paradox, yet it seems to be much more poorly explained within the usual interpretation of quantum mechanics.

Hidden Variables

The use of hidden variables was not a new idea in 1952, even in quantum mechanics. It was an understood idea already in statistical physics and thermodynamics. Here, the idea of atoms had been around for a very long time. Yet, in the early years it was recognized that we could not see these atoms. Nevertheless, each atom must have a position and momentum, and the other attributes, such as energy, that go along with these quantities. The fact that we could not see the atoms meant that we had no possible way of measuring the properties of the individual atom, but we did know the ensemble average values for the quantities for the gas of atoms. This we knew from thermodynamics, which was mainly an experimental science. And we knew it from statistical physics as developed by Boltzmann, Gibbs, and others. The fact that we could not see the atoms, and we could not measure the properties of the individual atom, did not mean that they didn't exist, although for sure Ernst Mach, and the positivists, tried hard to convince the scientific world that atoms didn't exist. As we pointed out in Chapter 1, Holton has described this objection as [25]: "what Ernst Mach was attacking when he objected to the notion of atoms ... was not the phenomenic [*sic*] hypothesis of the atom ... [but] the concept of fundamental submicroscopic discreteness as against continuity." This is an important insight that will affect the later assaults against the idea of hidden variables.

Later, with Bohr's model of the atom, there remained the question about the cause of the transitions between the various atomic levels. He carefully defined the levels themselves, but did not really address just what caused the transitions between the levels. That is, he was reluctant to actually introduce randomness in the problem. Einstein introduced such randomness with his theory of absorption and emission [26]. As we remarked in Chapter 2, Einstein noted that a molecule/atom can only exist in a series of states, each of which had a unique energy level, which followed Bohr's theories. He assumed that "a molecule may go from state Z_m to a state Z_n and emit radiation with an energy $E_m - E_n$" with a probability A_{mn} per unit time. He assumed that the reverse process would be characterized by an absorption process with probability B_{nm}, but then he realized that these were not proper *inverse* processes. The downward transition with probability A_{mn} was made in the absence of any radiation field, whereas the upward process with probability B_{nm} was made in the presence of an electromagnetic field, such as light. Hence, there must be an equivalent downward transition in which the molecule emits radiation *due to the presence of an electromagnetic field*. He said this emission process would be denoted by the probability per unit time B_{mn}. Hence, the B processes are stimulated by the presence of the electromagnetic field, while the A process could occur spontaneously. Thus, there was a causative effect on two of the processes, and this arose from the electromagnetic field, and the source of this causative action is hidden from the observer. On the other hand, the downward transition with probability A_{mn} was described as a random event even though a probability distribution could be assigned to it according to the Bose–Einstein distribution function (although this name came much later). Even then, Bohr remained reluctant to commit to the randomness. As Whitaker put it [27], "When asked about the mechanism of the transitions, he would reply that this was a part of the problem not yet fully resolved. This was almost certainly a means of keeping attention on the undeniably successful aspects of this theory, in particular agreement with experiment. Bohr did not want this success overshadowed by contentious debates on determinism." One could conclude, however, that if you could actually find a causative effect for this transition, it would imply a return to the view that the electron and its transition were real submicroscopic events, and this would upset the continuity that Bohr desired with his positivist philosophy.

Hidden variables also were intrinsic to de Broglie's quantum theory. As with the later developments by Bohm, de Broglie was able to show that the velocity of the particle was determined by the phase of the wave function. It was this reliance on the phase, instead of the amplitude of the wave, for the guidance of the "singularity" that represented the particle that provided the two-solution idea of particle *and* wave. In this regard, de Broglie thought that [28]

> the light-quantum theory cannot be regarded as satisfactory since it defines the energy of a light corpuscle by the relation $W = h\nu$ which contains a frequency ν.
>
> On the other hand the determination of the stable motions of the electrons in the atom involves whole numbers, and so far the only phenomena in which whole numbers were involved in physics were those of interference and of eigenvibrations. That suggested the idea to me that electrons themselves could not be represented as simple corpuscles either, but that a periodicity had also to be assigned to them too.

Thus, Cushing says [29], "The ψ-wave gave statistical information about the behavior of an ensemble of particles (or equivalently, probabilistic information about the behavior of a single particle whose initial location—the 'hidden variable'—is unknown or uncertain)."

Already by the 1927 Solvay conference, the idea of hidden variables, and their connection to reality, were an anathema to the Copenhagen–Göttingen crowd, and their adherents. In a sense, this was all about the Einstein–Bohr debates on the reality and completeness of quantum mechanics. If hidden variables were allowed into the discussion, then it was clear that the usual quantum mechanics was deficient in describing these contributions. In turn, this meant that quantum mechanics had to be incomplete in some manner. So, they had to go. The arguments were put in formal form by von Neuman in his book on the mathematical foundations of quantum mechanics [30]. Here, there appeared to be a proof that any type of hidden variable theory that agreed with usual quantum mechanics was impossible. Now, as Cushing points out [29],

> Von Neumann was able to produce a mathematical contradiction only by assuming that, for dispersion-free ensembles, the expectation value of the sum of two (noncommuting) operators is simply the sum of the expectation values of each separately.

Now, a dispersion-free ensemble is one which satisfies an eigenvalue equation, and two noncommuting operators cannot both satisfy eigenvalue equations. To do so would indicate that the two could be measured simultaneously, and this is not allowed for noncommuting operators. As Bell showed later [31],

> It was not the objective measurable predictions of quantum mechanics which ruled out hidden variables. It was the arbitrary assumption of a particular (and impossible) relation between the results of incompatible measurements either of which might have been made on a given occasion but only one of which can in fact be made.

Cushing points out [29], "Another way to parse this is that such a hidden-variables extension requires a more complete specification of the state of a microsystem than that possible with the quantum mechanical state vector ψ." To this point, it in fact does appear that the usual quantum mechanics is inadequate for this purpose and must be deemed to be incomplete. Bohm's approach brings the particle and wave viewpoint back into the picture as a viable approach to quantum mechanics. He specifically discounted von Neumann's proof [12]:

> [Von Neumann's] proof ... shows that the usual quantum-mechanical rules of calculating probabilities imply that there can be no "dissipationless states," i.e., states in which the values of all possible observables are simultaneously determined by physical parameters associated with the observed system.... With this conclusion, we are in agreement. However, in our suggested new interpretation of the theory, the so-called "observables" are ... not properties belonging to the observed system alone, but instead

> potentialities whose precise development depends just as much
> on the observing apparatus.... The final result is determined
> by hidden parameters in the ... measuring device as well as by
> hidden parameters in the observed electron.... Von Neumann's
> proof ... is therefore irrelevant here.

Now, we note that normal closed quantum systems have no
dissipation, because there is none in the Schrödinger equation.
Thus, to consider dissipation, von Neumann must apply to the
open system, which is likely in contact with the measurement
apparatus.

Since Bohm's work, there have been attempts to extend the
von Neumann theorem to do away with the possibility of hidden
variables. Arguments have been given by Jauch and Piron [32] to
this purpose. A more complete approach was also taken somewhat
earlier by Gleason [33]. It is not my purpose here to get into this
argument, especially as this work has been nicely summarized by
Bell [31]:

> These analyses leave the real question untouched. In fact ...
> these demonstrations required from the hypothetical dispersion
> free states, not only that appropriate ensembles thereof should
> have all measurable properties of quantum mechanical states, but
> certain other properties as well. These additional demands appear
> reasonable when results of measurement are loosely identified
> with properties of isolated systems. They are seen to be quite
> unreasonable when one remembers ... the impossibility of any
> sharp distinction between the behavior of atomic systems and the
> interaction with the measuring instruments.

Thus, it appears that there is actually no proof against hidden
variables that can actually be applied. Bell would try to rectify
that, as we will see in the next chapter. Yet, this did not stop
Bohr's colleagues from condemning attempts to move beyond the
Copenhagen–Göttingen religion. Rosenfeld, when presented with an
opportunity to speak following a talk by Bohm, could barely contain
his disdain, remarking [34],

> The critics have put forward the novel suggestion that an
> alternative "interpretation" could be found for the mathematical

formalism of quantum theory.... A physical theory is a consistent set of relations between certain concepts . . . but they are primarily the mental representation of concrete physical realities and as such must be defined in words, linking them . . . to *everyday experience*, i.e. to features of common observation....

The suggestion that such ambiguities might be circumvented by the introduction of "hidden parameters" is empty talk....

. . . we are here not faced with a matter of choice between two possible languages or two possible interpretations, but with a rational language intimately connected with the formalism . . . on one hand, . . . and with rather wild, metaphysical speculations . . . on the other.

Whitaker provides an insightful comment on this view [27]:

"Metaphysical" is a good slur-word for theories you don't like but can't prove wrong; it's rather like using the word "alcoholic" for "someone I don't like who drinks as much as I do."

References

1. E. Fermi, *Nature* **133**, 898 (1934).

2. I. Noddack, *Angew. Chem.* **47**, 0653 (1934).

3. O. Hahn and F. Strassman, *Naturwissen.* **27**, 11 (1939).

4. L. Meitner and O. R. Frisch, *Nature* **143**, 239 (1939); **143**, 276 (1939).

5. G. N. Flerov and K. A. Petrzhak, *J. Phys. USSR* **3**, 275 (1940).

6. F. Selleri, in American Institute of Physics Oral Histories: https://www.aip.org/history-programs/niels-bohr-library/oral-histories/28003-1

7. E. Schrödinger, letter to A. Einstein of 18 November 1950, reprinted in K. Przibram, ed., *Letters on Wave Mechanics* (Phil. Library, New York, 1967), p. 37.

8. A. Einstein, letter to E. Schrödinger of 22 December 1950, reprinted in K. Przibram, ed., *Letters on Wave Mechanics* (Phil. Library, New York, 1967), p. 39.

9. A. Einstein, *Dialectica* **2**, 320 (1948), reprinted in *The Born-Einstein Letters*, trans. by I. Born (Macmillan, London, 1971), p. 168.

10. D. Bohm, in American Institute of Physics Oral Histories: https://www.aip.org/history-programs/niels-bohr-library/oral-histories/4513

11. D. Bohm, in American Institute of Physics Oral Histories: https://www.aip.org/history-programs/niels-bohr-library/oral-histories/32977-5

12. D. Bohm, *Phys. Rev.* **85**, 166, 180 (1952).

13. E. Wigner, in *Proc. Intern. School of Physics "Enrico Fermi,"* course 49, *Foundations of Quantum Mechanics* (Academic, New York, 1971), pp. 1–18.

14. G. M. Prosperi, in Ref. 13, pp. 97–126.

15. D. Bohm, *Quantum Theory* (Prentice-Hall, New York, 1951).

16. L. de Broglie, *Comp. Rend.* **184**, 273 (1927).

17. E. Madelung, *Z. Phys.* **40**, 322 (1926).

18. E. H. Kennard, *Phys. Rev.* **31**, 876 (1928).

19. X. Oriols and J. Mompart, *Applied Bohmian Mechanics: From Nanoscale Systems to Cosmology* (Pan Sanford, Singapore, 2011).

20. G. I. Taylor, *Proc. Cambridge Philos. Soc.* **15**, 114 (1909).

21. A. Tonomura, J. Endo, T. Matsuda, T. Kawasaki, and H. Ezawa, *Am. J. Phys.* **57**, 117 (1989).

22. C. Philippidis, C. Dewdney, and B. J. Hiley, *Il Nuovo Cim.* **52B**, 15 (1979).

23. A. Einstein, B. Podolsky, and N. Rosen, *Phys. Rev.* **47**, 777 (1935).

24. D. Bohm and Y. Aharonov, *Phys. Rev.* **108**, 1070 (1957).

25. G. Holton, *Thematic Origins of Scientific Thought* (Harvard Univ. Press, Cambridge, 1973), p. 100.

26. A. Einstein, *Phys. Z.* **18**, 121 (1917); trans. in B. L. van der Waerden, *Sources of Quantum Mechanics* (Dover, Mineola NY, 1967).

27. A. Whitaker, *Einstein, Bohr and the Quantum Dilemma* (Cambridge Univ. Press, Cambridge, 1996), pp. 161–162.

28. L. de Broglie, Nobel Lecture, December 12, 1929, reprinted in *Nobel Lectures, Physics 1922–1941* (Elsevier Publishing, Amsterdam, 1965), pp. 244–256.

29. J. T. Cushing, *Quantum Mechanics: Historical Contingency and the Copenhagen Hegemony* (Univ. Chicago Press, Chicago, 1994), pp. 127–134.

30. J. von Neumann, *Mathematische Grundlagen der Quantenmechanik* (Springer, Berlin, 1932); trans. by R. T. Beyer, *Mathematical Foundations of Quantum Mechanics* (Princeton Univ. Press, Princeton, 1955).

31. J. S. Bell, *Rev. Mod. Phys.* **38**, 447 (1966).

32. J. M. Jauch and C. Piron, *Helv. Phys. Acta* **36**, 827 (1963).

33. A. M. Gleason, *J. Math. Mech.* **6**, 885 (1957).

34. L. Rosenfeld, in *Observation and Interpretation: A Symposium of Philosophers and Physicists*, ed. by S. Körner (Butterworth Sci. Publ., London, 1957), pp. 41–45, 107.

Chapter 8

Bell's Inequality

As we have seen, Einstein, with Podolsky and Rosen (EPR) [1], had set off a serious debate about the completeness of quantum mechanics and the resurgence of hidden variable viewpoints. Proposals for proofs against the possibility of hidden variables had been advanced both before and after EPR. Generally, it has been demonstrated by a number of authors that these proofs failed in their efforts. David Bohm had renewed both of the arguments illustrated by EPR when he proposed his causal interpretation of quantum mechanics [2]. Here, Bohm suggested that hidden variables were not only possible, but were necessary. Indeed, he pointed out that both position and momentum were, in fact, variables hidden from the observer, who had to rely only upon the probability function, the squared magnitude of the wave function. While Bohm's return to the wave *and* particle viewpoint connected in some sense to the ideas discussed by Bohr, they were certainly quite different from what has been called the Copenhagen–Göttingen interpretation of quantum mechanics. Heisenberg would argue that neither the position nor the momentum were meaningful variables, since neither was measurable; i.e., observable.

In the early 1960s, John Bell decided to look once again at the hidden variable problem in quantum mechanics, through an

The Copenhagen Conspiracy
David Ferry
Copyright © 2019 Pan Stanford Publishing Pte. Ltd.
ISBN 978-981-4774-75-8 (Hardcover), 978-1-351-20723-2 (eBook)
www.panstanford.com

examination of the EPR paradox [3]. Bell developed both a "theorem" and an inequality, but this may be overstating the case. In fact, he changed the argument. While his inequality supposedly ruled out *local* hidden variable theories, he was quick to point out that approaches such as that of Bohm were *nonlocal* theories.[a] This became a new controversy, and there would be questions as to the actual usefulness of the inequality, but it became quite clear that he was no fan of the Copenhagen–Göttingen interpretation of quantum mechanics. Baggott would write [4],

> I had never met Bell, nor heard him lecture, but in my reading of his scientific papers I have developed a great admiration for him and his work. I have especially admired his attempts to dismantle the orthodox Copenhagen interpretation of quantum theory, written with such tremendous style and obvious enjoyment.

In fact, it became known in the community that Bell, when asked his feelings about the Copenhagen interpretation, would paraphrase Shakespeare's Marcellus [5]: "Something is rotten in the state of Denmark." But this would not stop a significant number of physicists in claiming that quantum experiments, showing a violation of the inequality, demonstrated a "proof" for the theorem and the correctness of quantum mechanics, as expressed by the Copenhagen–Göttingen interpretation. Mermin would explain the contemporary viewpoints in the mid-1980s [6]:

> I think it is fair to say that . . . physicists would . . . [split] between the majority position of near indifference and the minority position of wild extravagance [to Bell's theorem] is an attitude I would characterize as balanced.
>
>   To moderate this point of view I would only add the observation that contemporary physicists come in two varieties. **Type 1** physicists are bothered by EPR and Bell's theorem. **Type 2** (the majority) are not, but one has to distinguish two subvarieties.

---

[a]The question about local versus nonlocal is often confusing. To me, the sense in which Bohm uses *nonlocal* refers to the fact that a particle going through one slit feels the quantum potential that arises from the entire wave function over all space. In this sense it is nonlocal in the way that a surface charge leads to a potential which is felt by a particle at some distance from the charge.

Type 2a physicists explain why they are not bothered. Their explanations tend either to miss the point entirely . . . or to contain physical assertions that can be shown to be false. **Type 2b** are not bothered and refuse to explain why. Their position is unassailable. (There is a subvariant of type 2b who say that Bohr straightened out [7] the whole business, but refuse to say how.)

Perhaps, it is justified to explain away such behavior on the part of the type 2 physicists by observing that Bohr himself had mostly stated that his opponents were wrong without ever really proving that they were in fact wrong. On the other hand, one must note that Mermin is generally regarded as being a supporter of the Copenhagen–Göttingen interpretation, and explanations he discounts are usually ones which question this interpretation.

Bell on Hidden Variables

John S. Bell was born in Belfast, Northern Ireland, and decided to become a scientist at an early age. He was educated at Queen's University in Belfast, and then went to Birmingham to pursue his doctorate in quantum field theory and nuclear physics. He pursued his career first in Britain, and then moved to CERN in Geneva, where he worked primarily on theoretical particle physics and accelerator design. Thus, the principles of quantum mechanics were more of a hobby to him rather than his principal occupation. Along the way, he collaborated on the translation of the series of Russian books on physics written by Landau and Lifshitz in the late 1950s and early 1960s. Whether a hobby or a principal interest, Bell raised major questions about the interpretation of quantum mechanics. He particularly was not happy with the prevailing view about the measurement and the supposed collapse of the wave function, writing [8],

> the observation of arbitrarily complicated observables, while not excluded in principle, is not possible in practice. It remains true that, whenever it is done, the wave packet reduction is not compatible with the linear Schrödinger equation. And yet at some not-well-specified time, such a reduction is supposed to occur [9]:

"... a measurement always causes the system to jump into an eigenstate of the dynamical variable that is measured..."

The continuing dispute about quantum measurement theory is not between people who disagree on the results of simple mathematical manipulations. Nor is it between people with different ideas about the actual practicality of measuring arbitrarily complicated observables. It is between people who view with different degrees of concern or complacency the following fact: so long as the wave packet reduction is an essential component, and so long as we do not know exactly when and how it takes over from the Schrödinger equation, we do not have an exact unambiguous formulation of our most fundamental physical theory.

The point he makes is that there is no transition defined which describes how the measurement moves the system from being described by Schrödinger to being a classical system described by the measurement apparatus. It cannot occur by decree. Even to say that the system becomes entangled with the measurement apparatus still requires a description of the interaction that causes such entanglement, and then a description of the ultimate "collapse" of the quantum wave function. So, it is not hard to understand how Bell came to be critical of the Copenhagen–Göttingen orthodoxy.

The question which Bell wished to address was precisely that of the possible existence of hidden variables. As he expressed it [10],

Once the incompleteness of the wave-function description is suspected, it can be conjectured that the seemingly random statistical fluctuations are determined by the extra "hidden" variables—"hidden" because at this stage we can only conjecture their existence and certainly cannot control them.

It was the Einstein–Podolsky–Rosen [1] paradox that brought most of the attention onto Bell and his equation. Here, he hypothesized his theorem, that is often expressed as no objective local theory can explain all of the quantum results. He stated that [11]

if [the theory] is local it will not agree with quantum mechanics, and if it agrees with quantum mechanics it will not be local. This is what the theorem says.

While he gave no proof, he felt that the theorem relied upon the inequality—if one was correct, the other would follow. The approach he took was the Bohm and Aharonov [12] version in which the two particles were created together in a manner such that their spins were oppositely directed, hence the total spin was zero. So, the two particles move apart, each with their own spin, and to some great distance. Let us call these two particles A and B. If the distance between the two particles is sufficiently large that the two measuring apparatuses cannot interfere with each other, then it may be assumed that the measurements will be independent of one another. The argument is based upon using, for example, Stern–Gerlach systems to measure the spin orientation of the two particles. Let us introduce, following Bell [3], two unit vectors $\mathbf{a}$ and $\mathbf{b}$ which point in the directions of the spin of the two particles A and B, respectively. Then, if we make a measurement of the σ_ξ component of the spin on particle A to find $\sigma_{\xi A} \cdot \mathbf{a} = 1$, we should suppose that the measurement at the second site should give $\sigma_{\xi B} \cdot \mathbf{a} = -1$. Hence, one may draw the conclusion that a measurement made of particle A would allow us to predict the value of the measurement on particle B. As Bell says [3], "this predetermination implies the possibility of a more complete specification of the state." However, *he adopts the view that the wave function does not describe a single measurement, but only an ensemble.* Then, one has to ask if the singlet wave function described by Bohm and Aharonov

$$\psi = \frac{1}{\sqrt{2}} [\psi_+(A)\psi_-(B) - \psi_-(A)\psi_+(B)], \qquad (8.1)$$

is a description of the entire quantum system or is merely a statement about our lack of knowledge of the quantum system. In fact, Bohm and Aharonov address the question in this way [12]:

> in the molecule, each component of the spin of particle A has, from the very beginning, a value opposite to that of the same component of B; and this relationship does not change when the atom disintegrates.
>
> In quantum theory, a difficulty arises, in the interpretation of the above experiment, because only one component of the spin of each particle can have a definite value at a given time. Thus, if the x component is definite, then the y and z components are indeterminate and we may regard them as in a kind of fluctuation.

In spite of the effective fluctuation described above, however, the quantum theory still implies that no matter which component of spin of A may be measured the same component of the spin of B will have a definite and opposite value when the measurement is over.

In the case of conjugate variables, the situation is somewhat different as they cannot both be measured, and the fluctuations are attributed to the measuring apparatus. But how would such fluctuations be accounted for with spin measurements that provide such exact results? The possible conclusion is that the wave function must be more complicated than the simple form assumed, perhaps by the presence of the hidden variables mentioned above.

Now, Bell decides to denote the set of hidden variables by the parameter λ. This parameter can represent one or more hidden variables. Then, the result of the above measurement of $\sigma_{\xi A} \cdot \mathbf{a}$ is determined by the value $\mathbf{a}$ and the values assigned to λ. Similarly, the result of the measurement $\sigma_{\xi B} \cdot \mathbf{b}$ is determined by the value $\mathbf{b}$ and the values assigned to λ. The expression of the state of the two particles is then summarized by the condition [3]

$$A(\mathbf{a}, \lambda) = \pm 1, \quad B(\mathbf{b}, \lambda) = \pm 1. \tag{8.2}$$

It may be presumed that the parameter λ is characterized by a probability distribution function $\rho(\lambda)$, and that the proper normalization of this is that the integral over all possible values of λ is unity. Then, the expectation value of the two values $\mathbf{a}$ and $\mathbf{b}$ are found in the two measurements is just

$$P(\mathbf{a}, \mathbf{b}) = \int d\lambda \rho(\lambda) A(\mathbf{a}, \lambda) B(\mathbf{b}, \lambda). \tag{8.3}$$

Unusually, this expectation value can be negative, given the values expressed in Eq. 8.2. And it can only reach -1 if we have the anti-correlation

$$B(\mathbf{b}, \lambda) = -A(\mathbf{b}, \lambda). \tag{8.4}$$

Bell's argument above that the two unit vectors point in the direction of the spins of the two particles lead to this result, as the measurement $\sigma_{\xi A} \cdot \mathbf{a} = 1$ must be matched with $\sigma_{\xi A} \cdot \mathbf{b} = -1$, since the directions of the spins are opposite and so must the two unit vectors. In this sense, the two unit vectors are antiparallel. Thus, with Eq. 8.4, we may rewrite Eq. 8.3 as

$$P(\mathbf{a}, \mathbf{b}) = - \int d\lambda \rho(\lambda) A(\mathbf{a}, \lambda) A(\mathbf{b}, \lambda). \tag{8.5}$$

At this point, Bell adds a third unit vector $\mathbf{c}$, without telling us if this is related to particle A or to particle B. Nevertheless, it follows that [3]

$$P(\mathbf{a}, \mathbf{b}) - P(\mathbf{a}, \mathbf{c}) = - \int d\lambda \rho(\lambda) \left[A(\mathbf{a}, \lambda) A(\mathbf{b}, \lambda) - A(\mathbf{a}, \lambda) A(\mathbf{c}, \lambda) \right]. \tag{8.6}$$

Equation 8.2 tells us that the product of either expectation with itself is 1, so we can insert this product, as a resolution of unity, between the last two probabilities in the square brackets, and then factor the common terms to yield

$$P(\mathbf{a}, \mathbf{b}) - P(\mathbf{a}, \mathbf{c}) = \int d\lambda \rho(\lambda) A(\mathbf{a}, \lambda) A(\mathbf{b}, \lambda) \left[A(\mathbf{b}, \lambda) A(\mathbf{c}, \lambda) - 1 \right]. \tag{8.7}$$

The magnitude of this may now be written as

$$P(\mathbf{a}, \mathbf{b}) - P(\mathbf{a}, \mathbf{c}) \leq \int d\lambda \rho(\lambda) \left| [1 - A(\mathbf{b}, \lambda) A(\mathbf{c}, \lambda)] \right|$$
$$\leq 1 - P(\mathbf{b}, \mathbf{c}), \tag{8.8}$$

where we have used Eq. 8.2 and the fact that the product of A's is -1, or

$$P(\mathbf{a}, \mathbf{b}) + P(\mathbf{b}, \mathbf{c}) - P(\mathbf{a}, \mathbf{c}) \leq 1. \tag{8.9}$$

The last form is generally considered to be Bell's inequality. Bell actually used the magnitudes, which makes it stronger, but this form is a standard inequality in probability theory that we will discuss below. It implies that the magnitude on the left-hand side of the expectation difference must be less than 1.

In contrast to the expectation found above, Bell has also considered the quantum expectation value for the two measurements, for which [3]

$$\left\langle \left(\sigma_{\xi A} \cdot \mathbf{a} \right) \left(\sigma_{\xi B} \cdot \mathbf{b} \right) \right\rangle = -\mathbf{a} \cdot \mathbf{b} = \cos \vartheta, \tag{8.10}$$

since the two are unit vectors and the last angle is the included angle between the unit vectors. Bell points out that, for a small difference between **b** and **c**, the left-hand side is linear in the magnitude of this difference. Thus, it is quite unlikely that $P(\mathbf{b}, \mathbf{c})$ will be stationary at its minimum value, nor is it possible for Eq. 8.9 to yield the quantum expectation value of Eq. 8.10. In fact, Bell asserts that there is not even an equivalence in any approximate expression. *The inescapable assumption is that if the quantum expectation value has values which will violate his inequality, one must rule out any local hidden variables.* By local, most interpretations are that Bell is ruling out any "action-at-a-distance" type of interactions or variables, but would accept that such forces could violate the inequality. A somewhat different form was obtained by Clauser, Horne, Shimony, and Holt, often referred to as CHSH [13], in which a fourth unit vector **d** was included. This form is expressed as

$$|P(\mathbf{a}, \mathbf{b}) - P(\mathbf{a}, \mathbf{d})| + |P(\mathbf{c}, \mathbf{b}) - P(\mathbf{c}, \mathbf{d})| \leq 2. \qquad (8.11)$$

What would experiments tell us? In fact, Alain Aspect, from the University of Paris, took up the challenge. Pairs of entangled photons are created and their polarizations are then measured in two separated experimental apparatuses, but the idea is that the polarization analyzers are not set until the last moment, and are sufficiently far from each other that light cannot have time to propagate between the two analyzers. The results of the experiments were a violation of Bell's inequality, more specifically of the CHSH inequality, from which they concluded that quantum mechanics cannot be a local theory [14]. It must include nonlocality. The experiments have been repeated in various forms several times in recent years, always with the same results. If quantum mechanics has hidden variables, they must be nonlocal in nature, if the results are to be accepted at face value. This will be an important discussion below.

Problems in Paradise

It is insightful to look at a simple example. Let us consider two coins. We presume that these coins are "fair" coins in that flipping them will produce an equal number of heads and tails, if we carry out a sufficiently large number of flips. To evaluate the flips, we use a Boolean logic of 1 and 0 for the outcomes, except we shift the axis. If the variable u has the value 1 or 0, we take the outcome as $2u - 1$, which now has values $+1$ and -1, respectively. These values correspond to those of Eq. 8.2 above. We now have to assign a value to heads and to tails and this is the purpose of the unit vectors **a** and **b**. For the first coin, we assume the unit vector **a** in the positive direction with heads giving $+1$ and tails -1. We can do a similar assignment for the second coin with the unit vector **b**. But now there is a problem. If we assign **b** to point in the positive direction for heads, there is a conflict with Eq. 8.4. According to this expression, if $A(\mathbf{a}, \lambda)$ returns $+1$ because the coin is heads, we are required to make an assumption. This is because if **b** also points in the same direction as **a**, then $A(\mathbf{b}, \lambda)$ also returns $+1$. To require that $B(\mathbf{b}, \lambda)$ return -1 means that the two coins must be anti-correlated or **b** is in the opposite direction of **a**. *While this may be a fair assumption for two entangled spins, it is certainly an unusual assumption for two coins.* And it is fair to say that others would point out that the sign really doesn't matter in the end, and perhaps that is true. But we know that the two coins can return four values as $++$, $+-$, $-+$, or $--$. We would normally expect $++$ to mean that both coins returned "heads," but Bell's spin-based assumption means that one coin was "heads" while the other was "tails."

The insertion of the additional unit vectors **c** and **d** (in the CHSH case) is significant. For example, to have both $A(\mathbf{a}, \lambda)$ and $A(\mathbf{c}, \lambda)$ means that two measurements on a single coin are being made. Quantum mechanically, this is unfair, because the first measurement is supposed to collapse the wave function, making the second measurement irrelevant. This is also true in the case of our two coins, because the first measurement fully determines whether the coin is heads or tails and the second measurement is thus irrelevant because it cannot change this result. This means that the right-hand side of Eq. 8.8 is identically 0. Such a result brings into question what the left-hand side of this inequality means. The answer given to us by the proponents of the usual quantum theory is that we really have to use sets of coins, and this of course fits with classical probability theory. For Eq. 8.8, we need three pairs of coins, while for CHSH we need four pairs of coins (or photons). To build statistics, we need to then generate an entire ensemble of pairs of coins. There is another issue here [15], for if λ represents something like the color of a particle or the composition of a particle (an analogy that we will use later), then it is possible that the settings of, say, **a** and **c** might encounter the same particle. On the other hand, if λ is independent of the various settings, then one expects to encounter approximately equal numbers of particles and so form a valid distribution. It is for this reason that Bell assumed the ensemble of spin pairs rather than discuss a single quantum system, but the above comment says that we have to be careful in what is meant by λ. In introducing the ensemble, Bell actually goes away from the EPR arguments, assuming that this won't be important. A philosopher such as John Norton would argue [16]:

> When we have a spread out quantum wave representing some particle, standard algorithms in the theory tell us what would happen were we to perform this measurement, or, instead of it, had we performed that measurement. Generally, the description of what would happen is expressed in terms of the probabilities of various outcomes. But, there is no difficulty in recovering the result. Thus, there seems to be no problem as far as quantum theory is concerned when EPR assert what would happen were this measurement or another incompatible measurement to be performed.

We have to remember, however, that it is exactly this "either . . . or" to which Bohr objected. In his view, the observation or measurement describes the quantum system, so the "either/or" actually is two different quantum system rather than a single one. This was the central point of his debate with Einstein. So, one has to be quite careful with such approaches, as we shall see.

The assumptions on **a** and **b** built into Eq. 8.4 suggest that the observers must communicate with each other to assure that the conditions for the inequality are met; e.g., the two unit vectors are antiparallel if one accepts that the sign in Eq. 8.10 is meaningful. This would be contrary to the assumption that the observers are far enough apart to avoid signaling. Indeed, the experimenters actually use random orientations of the polarizers. Again, the proponents assure us that the ensemble approach will yield the right results. They depend upon their ensemble averaging to produce the $\cos\theta$ given in Eq. 8.10 instead of the linear curve inherent in Eq. 8.8, and to give results which violate Eq. 8.9. This tells them that the interpretation of quantum mechanics, as put forward by the Copenhagen–Göttingen crowd, is correct. But this brings us back to the question of whether the probabilistic interpretation means that the wave function comes from an ensemble or may actually represent a single system. If the Schrödinger equation can represent a single system, then the problems of the inequality described here mean that it is irrelevant. Certainly, EPR assumed that they were dealing with a single system, and not an ensemble. Moreover, as mentioned, in the experiment, the observers set the polarizers at random angles, so that they are not parallel or antiparallel. This means that we cannot assure the condition Eq. 8.4 is actually met. In fact, if we set the angle randomly, we would expect that the right hand side is described by the $\cos\theta$, and Eq. 8.8 starts to look more and more like Eq. 8.10. So, how does a measurement tell us whether we have classical correlation or quantum entanglement? Again, we shall return to this question below.

Is the difference between the ensemble and a single system important? Of course. Let us consider again our two coins with the four possible outcomes listed above. Whether the two unit vectors are parallel or antiparallel, these four outcomes produce only two values, $+1$ and -1, that is the results are that the two coins are the same or opposite. So, a single pair of coins produces one of these two numbers. Let us now consider a very large number N of coin pairs. This should produce $N/2$ results of $+1$ and $N/2$ results of -1. When we average these results all together, we obtain the result 0. This ensemble average is quite different from the results obtained on a single pair of coins. Similarly, in Eq. 8.10, the possible values of the angle are 0 and π, which when averaged give $\pi/2$. The averaged result for the individual values, rather than the angle, is again 0. All this means is that the results of an ensemble average can produce an expectation value, which is not a possible result for any single system in the ensemble. It may be noted that the quantum expectation value yielded the same result as the classical average for our coins.

But there is a deeper problem, which lies in the desire to use a large number of so-called equivalent measurements and take ensemble averages. The problem was enunciated by Walter Philipp [17]:

> If something can be proven in such a simple way, no violation of that math is possible and no great science is behind it. The letters A, B, and C correspond to different experiments, and only two polarizer settings can be used for any given pair. Why should the A of the first term be the same as the A of the third, and why should the B's and C's in the different products be identical? The measurements are done for different entangled pairs! The fact that the settings could have been chosen differently for a given pair means nothing.

But this assertion, as with EPR, lies in the assumption that reality exists prior to the measurement, and the statement assumes that counterfactual thinking is possible. This is quite contrary to the basics of the Copenhagen–Göttingen interpretation of quantum theory. In this "usual" approach, counterfactual assumptions, the "either/or" type of arguments, are forbidden.

Moreover, Bell has assumed that the set of hidden variables, and more importantly the probability function of the hidden variables, is unchanging. Hess and Philipp began to worry about this assumption [18], because it has some implicit assumptions of its own. If the two entangled photons in the experiment are to be measured precisely, then it involves a simultaneous measurement. But, to Einstein, what is simultaneous in one frame of reference may not be simultaneous in a second frame of reference. If there was a delay so photons from two pairs were measured, the entire experiment was compromised. Hence, the set of hidden parameters were not arbitrary, but had constraints of their own. Then, the function A described at one time may well not be the same function at a later time.

Hess and Philipp proposed a thought experiment [17], in which one could simulate the EPR experiment. They would take two computers, one in Tenerife and one in La Palma (these are two islands in the Canary Islands, and their importance will be seen later). The choice is made so that no communications between the islands can occur during the window in time at which the measurements are made (the coincidence time). The computer in Tenerife has two software applications (apps) loaded so that it will choose a setting of **a** or **b**. The second computer also has two apps, but these will randomly select **b** or **c**. These apps start as soon as the arrival of the entangled pair occurs. The authors point out that any trigger signal which provides information on λ_i (where the subscript corresponds to a particular time window for which the experiment is performed) can be used. The internal clocks of the two computers have been synchronized beforehand. The actual detection of the signal will occur, for example, at t_i in Tenerife and t_i' in La Palma, both of which lie within the coincidence window. Now, suppose that in the 12th entangled pair, the app in Tenerife returns a $+1$ and a setting of **a**, while the app in La Palma returns a -1 and a setting of **b**. Then, we have $A(\mathbf{a}, \lambda_{12})A(\mathbf{b}, \lambda_{12}) = -1$. Then, they state [17], "If one performs many experiments with randomly chosen apps and a countable number of arbitrary λ_l, one always fulfills Bell's inequality." After this experiment, they performed a second one in which they introduced a time dependence, in which the actual apps were different at different measurement times, which means that the functions A will change with time. Now, they are no longer dealing with the strict Bell inequality, since the latter implies functions of independent variables. This turns out to have big consequences for the mathematics, and the results of the computer experiments now satisfied a new inequality, expressed as

$$A^i(\mathbf{a}, \lambda_i)A^{i'}(\mathbf{b}, \lambda_i) + A^{i+1}(\mathbf{a}, \lambda_{i+1})A^{i'+1}(\mathbf{c}, \lambda_{i+1})$$
$$- A^{i+2}(\mathbf{b}, \lambda_{i+2})A^{i'+2}(\mathbf{c}, \lambda_{i+2}) \leq 3. \qquad (8.12)$$

This is a more general result, and certainly the various experiments on entangled photons satisfied this expression. Hess points out that [17] "this would mean that outcomes could be different for different sequences of experiments. Actually, according to quantum theory, they indeed may be, and experimentally indeed are found to be different also. Different sequences of measurements may have different outcomes for the same setting. Only the long-term averages stay the same."

How could the different experiments, or sequences of experiments, yield such different inequalities? It turns out this is a real result in the theory of probability. There is a common paradox that illustrates such things. If one takes a single person at random, the chance that their birthday is on April 15 is something like 1/365 (we exclude leap year). So, if you took 23 people at random, you might expect that the likelihood of two people having identical birthdays is ~23/365. In fact, this probability is quite a bit larger, being about 50%. And if you take 75 people at random, it is 99.9% likely that two will have the same birthday.[b] This is termed the birthday paradox, and just serves to tell us that properly doing the mathematics of probability will sometimes lead to unexpected results. So, it will not be surprising to learn that Bell's inequality was known more than a century before Bell wrote it down. This is the point that Hess and Philipp make. More importantly, they point out that action at a distance is not necessary, and that violations of the Bell inequality do not prove that the Copenhagen–Göttingen interpretation of quantum mechanics, in which the measurement defines reality, is correct [18].

George Boole

George Boole was born in Lincoln, England, in 1815, the son of working parents. He completed only primary school, although he had some home schooling, particularly in languages. At the age of 16, he had to become the breadwinner for the family, and took a teaching position. Fortunately, he participated in the Mechanics Institute, an organization for adult education, and which came to Lincoln when Boole was 18. Here, he learned mathematics and unlocked his incredibly active mind. He was able to become sufficiently adept in this latter field, so much so that he became the first professor of mathematics at Queen's College, Cork, at the ripe old age of 33. He gained wide fame in mathematics, philosophy, and logic. In 1854, he published *The Laws of Thought*, which outlined his approach to binary logic and probability, among other things [19]. This work contains many of what became known as inequalities in probability, such as Bell's inequality. He focused on probability in a

---

[b]The correct approach may be found at https://betterexplained.com/articles/ understanding-the-birthday-paradox/

subsequent paper [20]. His contributions were so well recognized that he was elected a Fellow of the Royal Society and awarded honorary doctorates from Dublin and Oxford. It is probably fair to say that he was the father of modern probability theory, as well as the ubiquitous Boolean logic by which modern computers operate; i.e., the use of base 2 arithmetic with the symbols 0 and 1. Not only did he describe a series of inequalities, he also described the manner in which many of these could be violated in everyday life. And, in 1844, he described how mathematical operators may *not* commute [21]—81 years before Heisenberg and Schrödinger.

Boole's major concept of the 1854 book was to describe an algebra by which one could address the logic contained in the arguments of philosophy. As he remarked [19],

> The design of the following treatise is to investigate the funda-mental laws of those operations of the mind by which reasoning is performed; to give expression to them in the symbolical language of a Calculus, and upon this foundation to establish the science of Logic and construct its method; to make that method itself the basis of a general method for the application of the mathematical doctrine of Probabilities; and finally, to collect the various elements of truth brought to view in the course of these inquiries some probable intimations concerning the nature and constitution of the human mind.

In this approach, he mostly succeeded, and the algebraic logic method is still a major part of philosophy. He was not the first to approach this topic, as much is laid at the feet of Leibniz some two hundred years earlier, but Boole's logic and algebra were a major contributor to the overall effort. He also remarked that "*probability*, in its mathematical acceptance, has reference to the state of our knowledge of the circumstances under which an event may happen or fail" [19]. It is important to note here that he makes probability a discussion about *what may happen*, not what has happened.

One of the many examples Boole considers is that of three possible events, which are labeled x, y, and z. Now, for these events, he defines some probabilities: (i) the probability that x occurs or that it alone among the three fails to occur is p, (ii) the probability

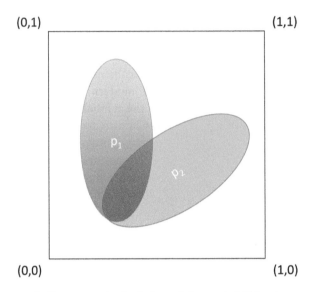

(0,1)   (1,1)

(0,0)   (1,0)

Figure 8.1 A diagrammatic depiction of the probabilities in expression 8.13.

that y occurs or that it alone among the three fails to occur is q, (iii) the probability that z occurs or that it alone among the three fails to occur is r. From these definitions, he finds a series of inequalities, one of which is

$$p + q - r \leq 1. \tag{8.13}$$

We wish to consider the case for which x and y are true, but z is false. In other words, the value of r relates to the case where both x and y are true, whereas the other two values are related to the case when x is true or when y is true. It is just the latter two in which we are interested, and let us investigate it with a diagram, Fig. 8.1. Let us suppose that the blue area is that area for which x is true, and which we designate by the probability p_1. In a similar manner, we designate the area for which y is true by the peach colored region, and we call this probability p_2. The total area of the square is unity. Now, the area for which *either* x or y is true is defined to be the *union* of the two areas, which is written as $p_1 \cup p_2$. On the other hand, there is an overlap of the two areas, where *both* x and y are true. This is called the *intersection* of the two areas, and is written as $p_{12} = p_1 \cap p_2$. The problem we have is that simply combining the two areas to form the union double counts the intersection area where the two overlap. We have to take out one of these double counted areas, so that the relationship 8.13 becomes

$$p_1 + p_2 - p_{12} \leq 1. \qquad (8.14)$$

This latter equation gives us the proper probability that either x or y is true (or equals 1, where failure is 0).

Let us now return to Bell's inequality and connect the current probabilities with his definitions of the events (for simplicity, we assume a steady state so that we don't have to worry about space-time issues). Thus, we define

$$p_1 = P(\mathbf{a}, \mathbf{b})$$
$$p_2 = P(\mathbf{b}, \mathbf{c})$$
$$p_{12} = P(\mathbf{a}, \mathbf{b})P(\mathbf{b}, \mathbf{c}) \sim A(\mathbf{a}, \lambda)A(\mathbf{b}, \lambda)A(\mathbf{b}, \lambda)A(\mathbf{c}, \lambda)$$
$$\sim A(\mathbf{a}, \lambda)A(\mathbf{c}, \lambda) \qquad (8.15)$$

where we have used the property 8.2. Hence, with these substitutions, we arrive at Eq. 8.9, Bell's inequality. (I suppose we should really refer to it as Boole's inequality.) Here, however, we have said nothing about hidden variables, nor have we discussed anything about the measurements. In fact, since this result dates from 1854, *it can have nothing to do with quantum mechanics.*

As mentioned above, Boole spent a great deal of time studying just how and why the inequalities could be violated. These have been nicely summarized by Pitowsky [22] as

(1) failure of randomness,
(2) measurement biases,
(3) no distribution, or
(4) mathematical oddities.

As an example of point 1, Pitowsky suggests considering a large bucket containing a vast number of balls of various colors and materials. Now, we draw a number of balls out of the bucket to make sample 1, and there are a fraction of balls which are red, denoted by p_1. We then take a second sample from the urn and discover a fraction of these that are made of wood, which we denote as p_2. Finally, we take a third sample and discover that a fraction of the balls are both red and wooden, which is denoted as p_{12}. As Pitowsky remarks "*If the samples are sufficiently large we expect that Boole's condition* [Eq. 8.14] *to obtain*" [22]. The reason for this assumption is a result of the law of large numbers, so that it is reasonable to assume that the number of balls in the bucket is so large that the fraction of red balls and wooden balls would obey this equation. Since the large quantity of balls in the bucket satisfies Eq. 8.14, we would be led to assume that our three samples should also satisfy the inequality. Nevertheless, if we obtain the values $p_1 = 0.6$, $p_2 = 0.7$, and $p_{12} = 0.28$, we have a violation of the inequality. Since we have the balls in our hands, points 2, 3, and 4 are not likely to be the cause of the failure. This just means that one or more of our samples is not sufficiently large to represent the actual population of the different color and composition of the balls in the bucket. We have violated the true randomness of the sample distributions. As Pitowsky summarizes the situation, "the probability that it [the failure] will occur decreases as the size of the samples grows. Thus, if we have samples of extremely large sizes, and the violation of at least one relevant condition persists, we should look for an alternative explanation" [22].

As we discussed above, Hess and Philipp [18] were of the opinion that the possible hidden variables could well have been time and experimental sensitivity which introduces an unsuspected measurement bias into the observables. Pitowsky explains this as follows: "Even when the samples are perfectly random we can still observe a violation of Boole's conditions which is due to a bias or 'disturbance' introduced by our method of experimentation" [22]. In our coin flipping experiment, one experimenter perhaps got tired, and her replacement used a different coin made of a magnetic material, which just may have introduced bias in the experiment. In the entangled photon experiments, there may have been an undiscovered trait of one of the experimenters, varying the polarizers, to make the change to a value dependent upon the previous measurement, but as remarked later, many-body interactions between the photon and the measuring system may have an effect that changes the system. Even when the adjustment is made mechanically, the bias may well reside within the computer controller that has been programmed by a human to set the new value for the polarizer. In a similar manner, problem 3 may just be an assumption on the part of the person analyzing the data. That is, we may jump to the conclusion that the probability distribution is a Maxwellian, even when the overall system is far from equilibrium. As Pitowsky puts it [22],

> Whenever we are faced with statistical data, in the form of relative frequencies, we tend to attribute the results to some pre-existing distribution of properties in a hypothetical population. This attribution is just a special case of the human habit to look for causal explanations, even in cases when only the effects are present.

An example of the mathematical oddity is the birthday paradox mentioned above.

The conclusion we draw from this discussion is that it is extremely over-reaching to assume that an experimental violation of Bell's inequality, even in a quantum experiment, can be used to rule out an assumption of hidden variables. As Hess and Philipp [18] correctly conclude, there are just too many possible places where hidden bias can creep into the experiment, for at least the four reasons mentioned above. Moreover, Hess and colleagues have reworked the inequalities in a proper form, and then the violations are not found [23], although not without some controversy. In addition, we cannot rule out other reasons for failure, such as those mentioned earlier in connection with Schrödinger's cat. In this latter case, and many others, one just cannot rule out Murphy's Law— what can go wrong will go wrong. Perhaps the best known example in in the last few years for an exceedingly high profile experiment to fail is the case of a measurement of the neutrino velocity [24]. In the experiment, announced in 2011 to the world, neutrinos were sent from CERN to the Gran Sasso lab in Italy, and it was found that they arrived about 60 ns earlier than expected. The conclusion was that the neutrinos were traveling faster than light, something Einstein would not have liked. In the end, however, an experimental error, due to an improperly functioning connector, caused the delay in the synchronizing signal reaching Gran Sasso. Murphy had struck again. And this was a simple time-of-flight experiment, and time synchronization had failed at the worst time. The Bell experiments are far more complicated.

CHSH

The experiments of Aspect were mentioned above. Subsequent experiments by a group from Vienna have received more attention as they clearly put the detection of the entangled photons relatively far apart [25]. The polarization entangled photons were generated atop the highest point on the island of La Palma, in the Canary Islands. One of the photons was detected on La Palma, while the second was transmitted to the island of Tenerife. The two photons were entangled so that one was in the horizontal polarization while the other was in the vertical polarization. This meant that the two polarizations were anti-correlated in any basis used for the detectors. Thus, two detectors at each site, with the appropriate beam splitters and mirrors, could measure these polarizations, and the data could be analyzed in terms of the CHSH inequality [13]. Naturally, since this was a quantum experiment on entangled photons, they recorded a violation of the inequality. We note that it is primarily this experiment to which Hess and Philipp suggested the possibility of bias as a cause. So, let us examine the inequality in a little greater depth.

To probe a little deeper into the inequality, we use a somewhat different notation, and follow the development of Ghirardi [26]. In the experiment, we are trying to measure the polarization of the two photons that are detected at the two distinct sites. We remind ourselves that prior to de Broglie and Schrödinger, the wave–particle picture referred to electromagnetic waves, and this will be significant. Certainly, the propagation of photons from La Palma to Tenerife involves electromagnetic waves. The polarization of an electromagnetic wave is defined as the direction of the electric field vector. A propagating plane wave, whether a continuous stream of photons or a single photon, has both an electric field and a magnetic field, which are oriented normal to one another. If the electric field points along the x axis, and the magnetic field points along the y axis, then the wave is propagating in the z direction. In this latter case, the polarization points along the x axis. These are the coordinates of the wave and may not be the coordinates of the observer, a point we return to shortly. The electric field of any polarized wave in an arbitrary polarization direction can still be decomposed into two orthogonal components, say E_V and E_H, such that the sum of the squares of these two components is equal to the square of the field of the polarized wave. In fact, this is the normal manner in which the polarization is measured; the wave is projected onto detectors on these two orthogonal axes, and the polarization angle inferred from the detected amplitudes with simple geometry [27–29]. With the principle axes of the detector used to designate the "vertical" and "horizontal" coordinates, we may write the vector electric field as (we take the magnitude of the electric field as 1 since it is the angles that are important)

$$\mathbf{E} = \cos \vartheta \mathbf{a}_V + \sin \vartheta \mathbf{a}_H. \qquad (8.16)$$

The angle here, as mentioned above, has two components: the relative orientation ϕ of the detector with respect to the laboratory coordinates and the actual angle ξ of the polarization with respect to these latter coordinates.

For our experiment, we have two waves, so we will need detectors at two sites, and we effectively measure two angles of the polarization which we will label A and B. From Bell's arguments, we can define a probability for coincident measurements as described in Eq. 8.3,

$$P(A, B) = E_A(\vartheta_A)E_B(\vartheta_B), \qquad (8.17)$$

where here the E's are the expectation values and not the electric field (we won't use the field value any further). Such a definition, in consideration of Bell's inequality, has been referred to as a form of local reality, in that it is assumed that each measurement can be treated as an independent quantity without any long-range nonlocal considerations. The probability, as used by Aspect et al. [14] or Zeilinger [25], is based upon the intensity of the signal, which arises from one-half the square magnitude of the correlation field, as (here, we have no difference in the vertical or horizontal coincidence signals), which will add an extra factor of $1/2$. Thus, the likelihood of a set of values depends upon the various angles. Thus, we may write these as

$$P(V, V) = P(H, H) = \frac{1}{2}p^2(A, B)$$

$$= \frac{1}{2}[\cos \vartheta_A \cos \vartheta_B + \sin \vartheta_A \sin \vartheta_B]^2$$

$$= \frac{1}{2}\cos^2(\vartheta_A - \vartheta_B) \qquad (8.18)$$

$$P(V, H) = P(H, V) = \frac{1}{2}[\sin \vartheta_A \cos \vartheta_B - \cos \vartheta_A \sin \vartheta_B]^2$$

$$= \frac{1}{2}\sin^2(\vartheta_A - \vartheta_B).$$

Aspect et al. [14] have given the probability of obtaining the correlation result in terms of Eqs. 8.5 and 8.6 with the expectation

$$E(A, B) = P(V, V) + P(H, H) - P(V, H) - P(H, V)$$
$$= \cos[2(\vartheta_A - \vartheta_B)]. \qquad (8.19)$$

We can now restate the CHSH inequality (11) with this present notation as

$$|E(A, B) - E(A, D)| + |E(C, B) - E(C, D)| \le 2. \qquad (8.20)$$

Now, Ghirardi provides us with the example of taking the angles as

$$\vartheta_A = 0°, \quad \vartheta_B = 22.5°, \quad \vartheta_C = 45°, \quad \vartheta_D = 67.5°. \qquad (8.21)$$

When these values are used in the estimator 8.20, for our entangled photons, we find that the left hand side becomes

$$\text{result} = |\cos(45) - \cos(135)| + |\cos(45) - \cos(45)|$$

$$= \left|\frac{1}{\sqrt{2}} + \frac{1}{\sqrt{2}}\right| + \left|\frac{1}{\sqrt{2}} + \frac{1}{\sqrt{2}}\right| = 2\sqrt{2} = 2.828 \qquad (8.22)$$

which clearly indicates a violation of the inequality. So, does the violation of this inequality indicate that there are no hidden variables and therefore the usual interpretation of quantum mechanics must be correct? In fact, it indicates nothing like that as one may expect from the above discussion of Boole's theory. First, the assumption of a fixed angle between the polarizations of the two photons is just the "lack of randomness" that is mentioned as item 1 in Pitowsky's discussion above. Second, we selected a very special set of angles, from which the inequality was known to be violated, thus violating the lack of "bias," item 2 in Pitowsky's list.

A fair question to ask at this point is just where in the above equations does the entanglement appear? The simple answer is that it does not appear anywhere in these equations. Rather, it is only an initial condition. The polarizers, and any beam-splitters or mirrors, satisfy Maxwell's equations and not quantum mechanics. They are entirely classical. All that is being measured is the polarization of the two photons, and the violation of the inequality arises at least because the relative angle between the two photons is not random, thus violating Pitowsky's condition 1. In fact, if we generate two photons from two linearly polarized lasers in such a manner that the two polarizations are perpendicular (as with the entangled photons) to one another, we should still get a violation of the inequality, because they still violate the randomness constraint. In fact, this has been demonstrated for arbitrary, but polarization-correlated, classical signals [30]. Indeed, a similar result has been obtained from correlated noise [31]. Along the same line, de Raedt and colleagues carried out simulations of the EPR experiments using photons [32]. Their model strictly satisfies Einstein's definitions of local causality. Yet, these optical experiments, which do not rely upon any concept of quantum theory, still reproduce the results expected from quantum theory for both experiments. By not relying upon quantum theory means that no entanglement is assumed in the photon pairs. Thus, their numerical experiment relies solely upon the propagation via Maxwell's equations and the normal electromagnetic properties of polarizers and mirrors. Yet, like other experiments, they can achieve a violation of the inequalities. The conclusion is that there is no magical action at a distance.

The likely fact is that the CHSH inequality is bound to fail for a variety of reasons, even due to the lack of a single distribution, point (3) above. As Krennikov describes it [33],

> Is it possible to construct the joint probability distribution ... for any triple of random variables?
> ... this question was asked and answered ... by Boole ... it is easy to show that the answer is negative.... If this Boole–Bell inequality is violated, then the joint probability distribution ... for such a system of three variables does not exist.

We emphasize that for mathematicians consideration of Bell-type inequalities did not provoke a revolutionary reconsideration of the laws of nature. The joint probability distribution does not exist because those observables cannot be measured simultaneously.

The issue of simultaneous measurement is an important one, because once one allows pairs of photons to be measured at different times, the arguments of Hess and Philipp become crucial. Here, we have to admit that the construction of Fig. 8.1 was a very special case, but it served adequately to frame the problem and illustrate that violations were possible.

Now, what about the question of bias? We should add that de Raedt and colleagues have looked at this with different simulations of the optical experiments, particularly those of Aspect [14], and find [34] that

> ... the passage time of the photons, which depends upon the relative angle between their polarization and the polarizer's direction, influences the correlations, demonstrating that the properties of the optical elements in the observation stations affect the correlations.

This suggests that the problem of bias, or of experimental error, is real, even if unintended, and points out the many-body corrections that might occur. It is certainly likely that experimenters, even the most careful ones, can unwittingly introduce bias into the measurements. For example, if they are well aware that they are only receiving correlated polarization sets, this may affect how they set the polarizers from one photon set to the next. To address this, and to isolate entanglement from classical correlation, the experiment must be more complicated. One approach is to create four large ensembles of photon pairs. The first ensemble would have the two photons created with precisely the same polarization, which may be regarded as either horizontal or vertical. This set of data is needed to ensure that the detection setup accurately measures the polarization correctly. The second ensemble of photons would have the angle between the two polarizations truly random, as in the numerical simulations [32]. The third set of photon pairs would have the two polarizations at right angles, but be from our pair of linearly polarized lasers with no entanglement between the two photons. The fourth ensemble is truly entangled photons. The signal sent at any instant of time is a pair of photons randomly chosen from one of the four ensembles. As it is important that the folks at the detectors do not know which ensemble is chosen, there can be no communication between the sender and the receiver other than the photons. The folks at the detectors merely record the time sequence of polarizer settings and detector signals. Only through the interlacing of the various photon sets can it be reasonably sure that no bias is built into the experiment. The data is then sent back to the source, so that it can be separated into the four classes of results, and then analyzed. It may well be that all four sets of data will violate the CHSH inequality, and this would be what Krennikov (and Boole and Pitowsky and . . .) would expect. But if the inequality is to be judged to have some credibility, then the data for ensemble 2 must satisfy it. The correlation built into the other three sets of data likely will cause a failure of each to satisfy the inequality, due to correlation between the photons in each pair. Doing so will demonstrate that there is no quantum magic in the experiment, and action at a distance does not explain the results. And there is always the possibility, however unlikely, that only set 4 violates the inequality, the cause of this result has to be found elsewhere. It is important to point out at this point that there are a variety of groups who doubt the relevance of these inequalities.

By triple of random variables, mentioned by Krennikov [33], one has to think once again about the experiment suggested by the Bell inequality (9). Here, one has three detectors so that three pairs of photons are required, with each pair carefully separated as to the detectors selected. Thus, Krennikov [33] points out that you just can't create a common probability distribution in this situation. Others, following an interpretation of quantum mechanics that differs from the Copenhagen–Göttingen theology, have suggested that this is not a proper test for quantum mechanics because it is not a consistent sequence of experiments. This latter group follows what is called the consistent histories approach, a view that is often referred to as GGHO after Griffiths, Gell-Mann, Hartle, and Omnès [35–37]. In the usual quantum mechanics, wave function collapse is considered the byproduct of measurements. In contrast, wave function collapse in the consistent histories approach is a mathematical procedure for calculating conditional probabilities during the quantum evolution [38, 39]. Probabilities are introduced as part of the axiomatic foundations of quantum theory, irrespective of measurements. We can think of a history as a sequence of alternatives at a specific set of times [40, 41]. These alternatives may be simple yes/no statements, but open the door for probabilities at each state. Thus, the states $|\psi_k\rangle$ at any point in time may each be characterized by a "probability" P_k for the possible states at that time. In the strictest interpretation, this probability is a projection operator which takes the entire Hilbert space to the allowed set of states, and so creates a trajectory in the spirit of Dirac [42] or Bohm [2]. By nature, these projection operators compress the phase space and introduce decoherence, which is why the consistent histories approach is often called the decoherence approach. But now questions about time and determinism become almost irrelevant, as the histories take on both of these attributes. In moving from one state to the next, it may be assumed that there is a time ordering in the states, with $\dots t_k < t_{k+1} < t_{k+2} \dots$, etc. In turn, the system evolves via the Hamiltonian, and a particular history can be written as [38]

$$H = (\dots P_{j+2} P_{j+1} P_{j \dots}), \tag{8.23}$$

where P_i is the probability that proposition i is true at t_i. In the Hilbert space, we then have to introduce projections onto the desired space, and propagators from one state to the next, which leads to

$$H = [\dots \hat{P}_{k+1} U(t_{k+1}, t_k) \hat{P}_{k \dots}], \tag{8.24}$$

where

$$U(t_2, t_1) = e^{-iH(t_2 - t_1)/\hbar}, \tag{8.25}$$

for a time-independent Hamiltonian. But the Bell experiments lack this clear history approach. They are random, with random detector locations. So, such an experiment is not admissible within the consistent histories approach and is therefore irrelevant, especially for any information on the "truth" of quantum mechanics.

Contradictions

As we have seen above, there are severe concerns over the correctness of the various inequalities, especially as they apply to determining the "truth" of quantum mechanics, but are they logically consistent? Let us consider the principal actor in this drama, and that is *entanglement*. The usual approach to an argument on the importance of entanglement is the spin wave function (Eq. 8.1). In this equation, it is assumed that we have a pair of particles or photons which are entangled. In this sense one of the particles has spin up and one has spin down, or one has horizontal polarization and one has vertical polarization, only we don't know which one has which. Thus, we write a possible wave function to express this lack of knowledge. The equation for the photons will be slightly different, but of the same form. What does this wave function mean? Again, the normal argument is that each individual state, say the one that describes particle 1, has two possible values, so that its Hilbert space is of dimension 2, which is often written as $\mathbb{C}^2$ ($\mathbb{C}$ for complex, because the wave function itself can be a complex quantity, and 2 for two dimensions). Then, the two spaces, one for each particle, are assumed to exist in a tensor product Hilbert space, which is the product of the two and written as $\mathbb{C}^2 \otimes \mathbb{C}^2$. The fact that the wave function cannot be decomposed into a simple product of one function in the first space and a second function in the second space means that the two particles are entangled. But this is not the only possibility. We could construct a four dimensional space, where the axes are determined by the four possible outcomes of the experiment; e.g., $++$, $--$, $-+$, or $+-$, or the set HH, HV, VH, HH for the photon polarizations. Such a Hilbert space is denoted, in the same format as used in the prior case, as $\mathbb{C}^4$. In this latter case, Eq. 8.1 is a simple vector in this Hilbert space and does not exhibit any magical entanglement; it is merely a vector. It is a general premise in quantum mechanics, and proved in many books, that the expectation value, or the eigenvalues, of an operator do not depend upon the basis set (or coordinates) chosen for the evaluation of that quantity [43]. But we have just seen that entanglement does depend upon a very specific basis set, the $\mathbb{C}^2 \otimes \mathbb{C}^2$. What does this mean? Well, quite generally, it has to imply that there is no operator which has a corresponding expectation value that tells us what entanglement is (and this difficulty of understanding entanglement may be why most books ignore it). Rather, we need to find a special, non-positive definite operator, which can serve as a *witness* [44]. What is meant is the following: we construct the density matrix from the wave function at different positions in space and compute the expectation value of our witness operator W. If this expectation value is positive, there is no entanglement, but if the expectation value is negative, then the density matrix contains entanglement. This does not imply that we know its value, or just what its properties are in the density matrix. That is, the entanglement is not directly measurable. However, many efficient techniques have arisen which infer the presence of entanglement and degree of it when it exists [45–48]. However, the existence of entanglement means that we need an exponential growth with time of the information required to describe the entangled state [49]. And then, there is the problem that the measurement entangles the quantum state with the classical state of the experimental system, which further degrades the quantum entanglement. The rub, however, is that if the entanglement is not directly measurable, it must be hidden from the observer, who must dig out the information to infer that it exists. This sounds exactly like a hidden variable. But if the entanglement is the cause of the violation of Bell's inequality (or the CHSH inequality), then there are no hidden variables. And if there are no hidden variables, there is no entanglement. Thus, we have a contradiction.

The contradiction leaves us with two possible choices. First, we can assume that entanglement is a real quantum fact of life. Erwin Schrödinger was a very good mathematician and physicist, and he asserted that entanglement may be the most important aspect of quantum mechanics [50]. So, it is probably safe to assume that entanglement is real and possible. Then we must draw the conclusion that Bell's inequality (and by extension that of CHSH) has absolutely nothing to do with quantum mechanics and cannot be taken to be a proof of the non-existence of hidden variables. The second choice is to conclude that Bell's inequality possesses a magical quantum property which rules out hidden variables, and therefore there cannot be any such thing as entanglement. The preponderance of the evidence discussed in this chapter suggests that the first choice is the correct one, and that Bell's inequality has no relevance to quantum mechanics. This unfortunately means that the beautiful experiments of the groups of Aspect and Zeilinger are irrelevant. While being beautiful experiments, they tell us nothing about the existence of hidden variables, although they certainly rely on one—the entanglement. But they can tell us nothing about the "truth" of Copenhagen–Göttingen, or do they?

In both sets of experiments, and in others where entanglement has been used, the interaction creates the entanglement. In EPR, the interaction of the two particles creates the entanglement between them. In the case of photons, the entanglement is introduced at the time of creation of the two photons, so exists from time zero, so to speak. Hence, both types of experiments have introduced reality to the quantum state at a time well before any measurement is made. The existence of the entanglement by itself tells us that reality exists before the measurement. This is exactly what Einstein claimed in EPR. Thus, if entanglement is important, we have to assume that this existence of reality prior to the measurement is also important to quantum mechanics. Moreover, both the entanglement and the particles, along with their wave function(s) have a temporal behavior that still satisfies Einstein's local causality. All of this is incompatible with, and a contradiction of, the assertions of Copenhagen–Göttingen. Neither the epistemology nor the ontology of quantum mechanics is that which Bohr, Born, and Heisenberg have thrust upon us. The fact that many interpretations differing from Copenhagen–Göttingen have arisen in the past few decades, perhaps is support of the view that an interpretation differing from that of Bohr, Born, and Heisenberg is necessary to understand quantum mechanics. Bohr's entire house of cards has to come down if we are to accept entanglement and its requirement of reality prior to the measurement. As Hess [17] has chosen to say, *Einstein was right*!

References

1. A. Einstein, B. Podolsky, and N. Rosen, *Phys. Rev.* **47**, 777 (1935).

2. D. Bohm, *Phys. Rev.* **85**, 166, 180 (1952).

3. J. S. Bell, *Physics* **1**, 195 (1964); reprinted in J. S. Bell, *Speakable and Unspeakable in Quantum Mechanics* (Cambridge Univ. Press, Cambridge, 1987), pp. 14–21.

4. J. Baggott, *The Meaning of Quantum Theory* (Oxford Univ. Press, Oxford, 1992), p. xi.

5. W. Shakespeare, *Hamlet*, Act 1, scene 4, 87–91.

6. N. D. Mermin, *Phys. Today* **38**(4), 38 (1985).

7. N. Bohr, *Phys. Rev.* **48**, 696 (1935).

8. J. S. Bell, *Helv. Phys. Acta* **48**, 93 (1970); reprinted in J. S. Bell, *Speakable and Unspeakable in Quantum Mechanics* (Cambridge Univ. Press, Cambridge, 1987), pp. 45–51.

9. P. A. M. Dirac, *Quantum Mechanics,* 4th ed. (Clarendon Press, Oxford, 1958), Sec. 10.

10. J. S. Bell, in *Foundations of Quantum Mechanics: Proceedings of the International School of Physics Enrico Fermi, Course 49* (Academic Press, New York, 1971), pp. 171–181.

11. J. S. Bell, *Epistem. Lett.*, November 1975; reproduced in J. S. Bell, *Speakable and Unspeakable in Quantum Mechanics* (Cambridge Univ. Press, Cambridge, 1987), pp. 63–66.

12. D. Bohm and Y. Aharonov, *Phys. Rev.* **108**, 1070 (1957).

13. J. F. Clauser, M. A. Horne, A. Shimony, and R. A. Holt, *Phys. Rev. Lett.* **23**, 880 (1969). This particular form is from G. Ghirardi, *Sneaking a Look at God's Cards* (Princeton Univ. Press, Princeton, 2005).

14. A. Aspect, J. Dalibard, and G. Roger, *Phys. Rev. Lett.* **49**, 1804 (1982); A. Aspect and P. Grangier, in *NATO ASI, Series B*, **47**, 201 (1995).

15. K. Hess, private communication.

16. J. D. Norton, unpublished, http://www.pitt.edu/~jdnorton/teaching/HPS_0410/chapters/quantum_theory_completeness/index.html.

17. K. Hess, *Einstein was Right* (Pan Stanford Publishing, Singapore, 2015).

18. K. Hess and W. Philipp, *Proc. Natl. Acad. Sci.* **98**, 14224 (2001); **98**, 14228 (2001).

19. G. Boole, *An Investigation of The Laws of Thought: The Mathematical Theories of Logic and Probability* (Walton and Maberly, London, 1854).

The book has been reprinted by several publishers, but is available online at archive.org (https://archive.org/details/investigationofl00 boolrich) and from Project Gutenberg (www.gutenberg.org/files/ 15114/15114-pdf.pdf).

20. G. Boole, *Philos. Trans. R. Soc. London* **152**, 225 (1862).

21. G. Boole, *Philos. Trans. R. Soc. London* **134**, 225 (1844).

22. I. Pitowsky, *Brit. J. Phil. Sci.* **45**, 95 (1994).

23. H. de Raedt, K. Hess, and K. Michielsen, *J. Comp. Theor. Nanosci.* **8**, 1011 (2011).

24. T. Adam, et al., *J. High Energy Phys.* **2010**(12), art. 093.

25. R. Ursin, F. Tiefenbacher, T. Schmitt-Manderbach, H. Weier, T. Scheidl, M. Lindenthal, B. Blauensteiner, T. Jennewein, J. Perdigues, P. Trojek, B. Omer, M. Fürst, M. Meyenburg, J. Rarity, Z. Sodnik, C. Barbieri, H. Weinfurter, and A. Zeilinger, *Nat. Phys.* **3**, 481 (2007).

26. G. Ghirardi, *Sneaking a Look at God's Cards* (Princeton Univ. Press, Princeton, 2005).

27. J. P. Craig, R. F. Gribble, and A. A. Dougal, *Rev. Sci. Instru.* **35**, 1501 (1964).

28. A. A. Dougal, J. P. Craig, and R. F. Gribble, *Phys. Rev. Lett.* **13**, 156 (1964).

29. H. Heinrich, *Phys. Lett.* **32A**, 331 (1970).

30. D. K. Ferry, *Fluc. Noise Lett.* **9**, 395 (2010).

31. D. K. Ferry and L. B. Kish, *Fluc. Noise Lett.* **9**, 423 (2010).

32. S. Zhao, H. de Raedt, and K. Michielsen, *Found. Phys.* **38**, 323 (2008).

33. A. Yu. Krennikov, *Theor. Math. Phys.* **157**, 1448 (2008).

34. H. de Raedt, K. de Raedt, K. Michielsen, K. Keimpema, and S. Miyashita, *J. Phys. Soc. Jpn.* **76**, 104005 (2007).

35. R. B. Griffiths, *J. Stat. Phys.* **36**, 219 (1984).

36. M. Gell-Mann and J. B. Hartle, in *Complexity, Entropy and the Physics of Information*, ed. by W. H. Zurek (Addison-Wesley, Redwood City, 1990).

37. R. Omnès, *J. Stat. Phys.* **53**, 893 (1988).

38. R. B. Griffiths, *Consistent Quantum Theory* (Cambridge Univ. Press, Cambridge, 2003).

39. M. R. Sakardei, *Can. J. Phys.* **82**, 1 (2004).

40. J. B. Hartle, *Phys. Scr. T* **76**, 67 (1998).

41. R. Omnès, *The Interpretation of Quantum Mechanics* (Princeton Univ. Press, Princeton, 1996).

42. P. A. M. Dirac, *Rev. Mod. Phys.* **17**, 195 (1945).

43. See, e.g., E. Merzbacher, *Quantum Mechanics*, 2nd ed. (John Wiley, New York, 1970).

44. M. Mintert, C. Viviescas, and A. Buchleitner, in *Entanglement and Decoherence*, ed. by A. Buchleitner, C. Viviescas, and M. Tiersch (Springer-Verlag, Berlin, 2009), pp. 61–86.

45. S. L. Braunstein, *Phys. Lett. A* **219**, 169 (1996).

46. M. B. Plenio and S. Virmani, *Quantum Inf. Comput.* **7**, 1 (2007).

47. R. Horodecki, P. Horodecki, M. Horodecki, and K. Horodecki, *Rev. Mod. Phys.* **81**, 865 (2009).

48. H. Wunderlich, S. Virmani, and M. B. Plenio, *New J. Phys.* **12**, 083026 (2010).

49. A. Ekert and R. Jozsa, *Philos. Trans. R. Soc. London, A* **356**, 1769 (1998).

50. E. Schrödinger, *Naturwissen.* **23**, 807, 823, 844 (1935).

Chapter 9

Measurement

When we think of any of the various "paradoxes" that have been put forward in quantum mechanics, it is clear that measurement is a key issue. In the last chapter, measurement lay at the heart of the discussion over the Einstein–Podolsky–Rosen paradox and over the existence of hidden variables. Indeed, almost from the beginning, the issue of measurement has been part of the discussion of quantum mechanics. In addition, this has been complicated by the argument over reality and when and where it arises in the quantum system. All of this points to a conundrum—what is a measurement? If we are to say that a measurement is defined by the experimental apparatus, as I would suspect Bohr would have us believe, then how are we supposed to express the possible outcomes? How does the experiment interact with the quantum system? What is meant by *collapse* of the wave function? These are not separate questions, but are strongly interrelated with each other. Moreover, they are also connected with what we believe the wave function to be. It is hard to not ask yourself, "What does this all mean?" Perhaps before proceeding, we should remind ourselves about the character of physical law and science, as expressed by Richard Feynman [1]:

The Copenhagen Conspiracy
David Ferry
Copyright © 2019 Pan Stanford Publishing Pte. Ltd.
ISBN 978-981-4774-75-8 (Hardcover), 978-1-351-20723-2 (eBook)
www.panstanford.com

What is necessary "for the very existence of science," and what the characteristics of nature are, are not to be determined by pompous preconditions, they are determined always by the material with which we work, by nature herself. We look, and we see what we find, and we cannot say ahead of time successfully what it is going to look like. The most reasonable possibilities often turn out not to be the situation. If science is to progress, what we need is the ability to experiment, honesty in reporting results— the result must be reported without somebody saying what they would like the results to have been—and finally . . . the intelligence to interpret the results. An important point about this intelligence is that it should not be sure ahead of time what it must be. It can be prejudiced, and say "That is very unlikely; I don't like that." Prejudice is different from absolute certainty. I do not mean absolute prejudice—just bias. As long as you are only biased it does not make any difference, because if your bias is wrong a perpetual accumulation of experiments will perpetually annoy you until they cannot be disregarded any longer. They can only be disregarded if you are absolutely sure ahead of time of some precondition that science has to have. In fact it is necessary for the very existence of science that minds exist which do not allow that nature must satisfy some preconceived conditions.

In some sense, this is a cry for acceptance of the fact that serendipity is almost a requirement for the advancement of science. Others may say that nature is a very perfidious individual—shocking us just when we think we understand the object of our experiments. However, the unexpected result arising from an experiment cannot be construed to mean that reality is defined by the experiments. Instead, we must accompany the unexpected result with the realization that the theory was far from adequate.

Scientists had been studying and using the Hall effect [2] for more than a century as a standard characterization tool for semiconductors and metals. It could be said that, after this century of work, we had a good understanding of the Hall effect. Yet, when it was carefully studied at low temperatures in a quasi-two-dimensional system, we learned just how wrong we were. The discovery of the quantum Hall effect [3] radically changed our understanding of condensed matter physics as well as the field of metrology [4], but the experiment did not give us what the theory did, and that was the recognition of the importance of topology in condensed matter physics. Today, we are searching for a wide range of new materials which possess topologically protected states [5], such as the quantum Hall effect. Needless to say, quantum theory is fundamental in this search, and Schrödinger's equation remains an essential basis for the search. But we have gone a long way from the simple forms of some of the wave functions.

Probability

Max Born set the table for the interpretation of quantum mechanics when he hypothesized, in a brief communication on scattering of the wave function, that the magnitude squared of the wave function was the probability [6]. In his interpretation, this probability was the likelihood of finding the "particle," described by the wave function, at a position x. This is a break from the understanding preferred by Schrödinger, who preferred to think of the wave function being related to a charge density, or particle density in the many-body case. As Beller put it, the Born interpretation [7] "signifies abandonment of determinism, and introduces a new kind of reality, abstract and 'ghost-like'." The literature is full of various views of how and why Born came to his conclusions about this probability. The most likely is that of Beller [7], who suggests that Born was merely trying to describe a mechanism for Bohr's quantum jumps, and this seems likely enough. But, as we have seen, Schrödinger had no love of these quantum jumps and so likely felt no love for the probability interpretation either.

Even if we accept the view of Born, it does not solve many problems. In fact, it cannot explain many new concepts in quantum mechanics. This is because the probability uses the squared magnitude of the wave function, which removes any dependence upon the phase of the wave function. Yet, a great many new concepts and ideas depend upon this phase, such as the Aharonov–Bohm effect, where interference can be observed when the wave function, or the particles if one prefers, pass around two different paths which enclose a magnetic field [8]. We now know that this phase is more important and is referred to as a type of geometrical phase in the wave function, commonly called today the Berry phase [9]. And we know that a gauge transformation affects the phase of the wave function. Hence, the introduction of a Chern–Simons gauge transformation [10] leads to the recognition of the existence of the composite fermion, a new quasi-particle found at very high magnetic fields in the quantum Hall effect [11]. It appears therefore that the concept of probability expressed by Born is difficult to exercise in many areas of modern physics. In fact, it may also be very difficult to introduce properly more usual experiments such as those investigating the EPR effect.

For sure, the wave function satisfies the time-dependent Schrödinger equation. And if we are considering particles which are moving, we have to accept that Born's probability is also a time-dependent function. But, now, at which time are we to describe the probability? Consider, for example, either the EPR situation with the generation of two particles, or the coherent photon equivalent experiment, at an initial point in time. It is at this time that the quantum system may be said to be initiated along with its wave function, but if we move along to the point of measurement, is the probability description the same one that arose at the instant of creation? If we take the form of the two entangled photons of the last chapter and let the wave function describe the polarization of the two photons, we may write it as[a]

$$\psi(x, s, t) = \frac{1}{\sqrt{2}} [\psi_1(x, H, t)\,\psi_2(x, V, t) - \psi_1(x, V, t)\,\psi_2(x, H, t)].$$

$$(9.1)$$

Certainly it can be reasonably argued that this does describe the two photons at their moment of creation, as we are not sure which photon is horizontally polarized and which is vertically polarized. But, at the detector end of the experiment, are we sure that this is still a viable description of the state of the quantum system? Or, do we merely use this as a description of our ignorance of the photons arriving at the detectors? Perhaps we should describe this state as one of arrogant ignorance, since we persist in demanding that the

---

[a]The terms vertical (V) and horizontal (H) are for photon polarization. For spin, we would normally use up or down as the two designations.

two photons remain entangled, without any major influence of the measurement equipment and its arrangement in space and time.

On March 24, 2001, the Arizona Diamondbacks were playing the San Francisco Giants in an evening preseason baseball game. On the mound for the Diamondbacks was the future hall-of-famer Randy Johnson. Somewhere in the world of statistics, one could find out just how many strikes and how many balls were thrown by Johnson, but the decision as to whether a given pitch was a strike or a ball was always made by the umpire who observed the pitch as it crossed home plate (this scenario would have been loved by Bohr). So, one could give a probability that a given pitch would be a strike, but the judgement would not be made by the pitcher. Hence, Born would be happy, but on this night, as Johnson unleashed one of his famous fast balls, a hefty bird flew in front of home plate, and was demolished by the ball. The bird disappeared in a cloud of feathers, and the ball never reached home plate. Now, why should we care about this? Probably because this possibility did not enter into the probabilities for the outcome of the pitch. We saw in the last chapter that George Boole told us that the probability tells us about what may happen, not what actually happened. Nevertheless, there is nothing in Born's description of the probability that would account for the actual outcome of this scenario.

There are believed to be about 436 different species of birds in the Canary Islands. A number of these, such as hawks, eagles, and harriers, are known to be relatively high flying birds. In the previous chapter, we discussed the Viennese experiments on entangled photons in the Canary Islands as a test of Bell's inequality. So, how can we be absolutely sure that none of the photons has met one of these birds during its flight between the ends of the optical experiment? Do we cast the possibility of a bird encountering the photon as part of the quantum system or part of the measuring apparatus itself? To be sure, the birds are only a metaphor for a wide variety of scattering centers that can exist in the atmosphere. The Canary Islands are in an oceanic climate, and water vapor is known to absorb light throughout the electromagnetic spectrum. With birds, and water vapor, and who knows what else, how can we completely assert that the photons remain entangled; in fact it is likely that some 4–5% of the photons will not make it to the detectors, and if some do undergo various collisions, with possible decoherence or even disappearance, we have introduced a time dependence in the process; worse, this is likely to be a random time variation. Such absorption or scattering processes are certainly hidden variables in the sense of the experiment, especially if we constrain ourselves to Eq. 9.1. The introduction of a possible time dependence to the system and to the hidden variables by Hess and Philipp [12] (who introduce a dependence on many-body interactions, with the local equipment that depends on time and equipment settings) is demanded by the scenario of the experiments. It is easy to understand the importance of time synchronization in this experiment if any sense is to be made from the results.

The Aharonov–Bohm effect was mentioned above. Let us return to this effect and an experiment demonstrating its measurement. The structure, and several measurements, are shown in Fig. 9.1 [13]. The material is graphene, which is a monolayer of carbon atoms in a honeycomb structure. The device is patterned by electron beam lithography and the structure is then defined by etching which removes selected areas of the graphene. Hence, the electrons exist just in the darker grey areas shown in the inset of panel (a). The current flows between the left and right "contact" areas, and the voltage is measured across this region. If we just consider the electrons flowing as individual quantum particles, we will not see any interference effect. However, if we have a single wave function extending through the entire structure, we can talk about the wave in the top side of the circular ring as being described by $Ae^{i\varphi_1}$ while that in the bottom side of the ring may be written as $Be^{i\varphi_2}$. We may now write the probability in the ring as

$$
\begin{aligned}
P_{\text{ring}} &= \left| Ae^{i\varphi_1} + Be^{i\varphi_2} \right|^2 \\
&= A^2 + B^2 + 2AB\cos(\varphi_1 - \varphi_2).
\end{aligned}
\tag{9.2}
$$

Normally, there should be no phase difference between the two paths. However, if we pass a magnetic field through the center of the ring, then the resulting vector potential affects the phase. One path gains phase and the other path loses phase. The result is an oscillation in the conductance between the top and bottom of the structure which oscillates in this phase difference. Obviously, the

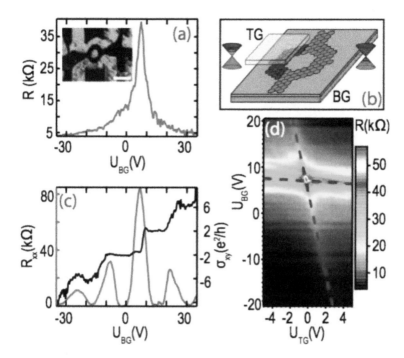

Figure 9.1 (a) Four probe resistance measurements over the ring versus backgate voltage. The inset shows an atomic force microscope picture of the sample. (b) Schematic picture of the graphene ring with different charge carriers in the ring. (c) Longitudinal resistance and Hall conductivity versus backgate voltage with a magnetic field of 13 T applied. (d) Resistance measurements for different topgate and backgate voltages showing two charge neutrality lines. R in (a) is a two-terminal measurement, while R_{xx} in (c) is a four-terminal measurement. Reprinted from Smirnov et al. [13] with the permission of AIP Publishing.

conductance is a maximum when

$$\varphi_1 - \varphi_2 = 2n\pi, \qquad (9.3)$$

where n is any integer (including zero). This oscillation in transmission through the ring is the Aharonov–Bohm effect [8]. The key factor is that the phase varies as a function of the magnetic flux coupled through the ring such that

$$\Delta\varphi = 2\pi \frac{eBS}{h}, \qquad (9.4)$$

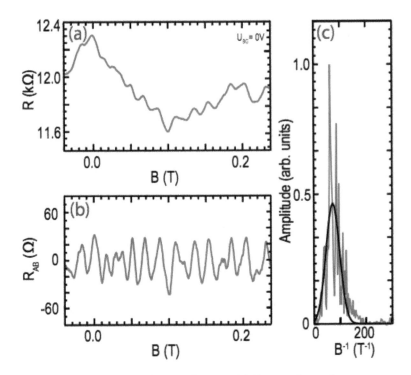

Figure 9.2   Aharonov–Bohm oscillations: (a) Four probe resistance measurements over the ring versus a perpendicular magnetic field at a constant backgate voltage $U_{BG} = 0$ V. (b) Same oscillations with subtracted background resistance. The period of the oscillations is 16 mT. (c) Fourier spectrum of the oscillations (red) and a Gaussian fit (black). Reprinted from Smirnov et al. [13] with the permission of AIP Publishing.

where B is the magnetic field, S is the area of the ring, e is the electronic charge, and h is Planck's constant. It is important to realize that the ratio h/e is the quantum unit of flux. In Fig. 9.2, the measured signal is shown. The raw data is given in panel (a), and the extracted oscillatory part in panel (b). The fine structure in (b) is likely the result of disorder in the material. Panel (c) is the Fourier transform showing the single frequency, which is related to e/h. The Fourier transform tells us that the signal is basically a single frequency governed by the phase relation 9.4.

The interesting point about this structure is that the oscillatory behavior arises from the ring structure at the center of the device, but conductance is a global behavior. We know that the semiconductor graphene is a quantum object, as this was established years earlier by Clinton Davisson and George Thomson (discussed in Chapter 2) through the observation of diffraction patterns from electrons striking metal films [14, 15]. So where does the measurement occur. The extension of the graphene from the actual ring means that the quantum behavior also extends from the ring. It can't be the contacts to the graphene or the wires from these contacts, as their behaviors is also governed by the quantum mechanics of the metallic material. The electronic measuring equipment is composed of wires and semiconductor devices, so it is basically quantum at the core of its operation, although we assume it to be classical. The same is true of any microchips in the device, and of the light-emitting diodes in the display. The point is that the semiconductor is basically governed by the wave mechanics and periodicity of the atomic arrangement to provide the quantum mechanical band structure which defines its properties. This structure is largely unaffected by measurements, whether these be mechanical, optical, or electrical. It is probably very fair to say that the semiconductor has a quantum mechanical reality that exists without any resort to measurement. Certainly, it isn't determined at the point at which the display is sensed by an observer. Since the band structure and properties persist through a great variety of measurements largely unchanged, it is apparent that the wave function of the quantum solid is not being "collapsed" by these measurements. Let us examine this a bit further.

Wave Function Collapse

The idea of wave function collapse is thought to have originated with Heisenberg's first paper where he points out that "the [bound] oscillating electron behaves like a free electron when acted upon by light of much higher frequency than any eigenfrequency of the [atom]" [16]. But it is in the uncertainty paper that he discusses the *reduction* of the wave packet. Here, he talks about multiple measurements of the position of an electron [17]:

> the classical "orbit" . . . comes into being only when we observe it. . . . Let a new determination of position be made after some time with the same precision. Its result . . . can be predicted only statistically. . . . The situation would be no different in classical theory, for there too the result of the second position measurement would be predictable only statistically because of the uncertainty in the first measurement. . . . Every position determination reduces the wave packet back to its original extension.

We note here that he was of the opinion that quantum mechanics would give quite similar results to classical theory after the measurement had been made, but does not require the collapse of the wave function during the measurement. Indeed, he offers the view that the wave function continues to exist during and after the measurement. This is an important view as it differs from many subsequent comments on the basis of quantum mechanics.

A somewhat different view of collapse is thought to have begun with comments Dirac made during the general discussion at the end of the 1927 Solvay conference. Here, he commented [18]:

> The state of the world at any time is describable by a wave function ψ. . . . It may happen that at a certain time . . . [it] can be expanded in the form . . . [of a series expansion over a basis set of functions, where each of the basis functions] are wave functions of such a nature that they cannot interfere with one another at any subsequent time . . . then the world at time later than [this] will be described not by ψ but by one of the [basis functions]. The particular [one] that it shall be must be regarded as chosen by nature.

It is noted that Heisenberg would have preferred selection by the observer rather than by nature, but the idea is that total wave function has collapsed into a single one of the possible basis states. Dirac is not explicit that this is a result of the measurement, but this is likely the underlying assumption. A later interpretation assigns the reduction to what is called the projection postulate—the state is selected by a projection of the total wave function onto the single state. Dirac goes on to express it slightly differently in his book [19]:

> When we measure a real dynamical variable . . ., the disturbance involved in the act of measurement causes a jump in the state of the dynamical system. From physical continuity, if we make a second measurement of the same dynamical variable . . ., the result . . . must be the same as that of the first.

Here, Dirac again postulates that the entire wave function is projected onto an eigenstate of the particular variable in question. Hence, the result of the same measurement made a second time cannot change. Von Neumann formalized this projection procedure, but noted that the state selected may be one from the measuring apparatus [20]. Von Neumann also pointed out a crucial issue, in which evolution of the wave function according to the Schrödinger equation was thermodynamically *reversible*, while projection of the wave function during the measurement was thermodynamically *irreversible* [21]. This is seen to be a crucial concept, that the act of measurement introduced irreversibility and collapse into the entire quantum system. Now, Bohm thought that there was no actual collapse of the wave function, but one measures only a "reduced" wave function [22], which suggests that the measurement picks out only a part of the total system. The difference in views covers a wide spread of interpretations.

We remarked above on the semiconductor graphene, which is a single layer of carbon atoms arranged in a honeycomb lattice. The electrons in the material are waves [14, 15], and the electronic structure is found from solving Schrödinger's equation in the material. Here, the wave function is subject to a periodic array of atomic potentials, and this periodicity creates special conditions for the solutions of Schrödinger's equation. The periodic array of atoms acts like a filter on the waves, in that only selected frequencies (energies) of waves are allowed. These allowed sets of energies are called bands, and the forbidden energies are called gaps. The electronic structure for graphene was first solved by Wallace [23], and it is somewhat peculiar. Normally, we are interested in the topmost occupied energy band and the next higher-lying band. The topmost occupied band is termed the valence band as it is formed from the bonding interaction of the outer shell electrons of the atoms. The next highest band, the conduction band, is normally empty in semiconductors and arises from the antibonding interactions of the outer shell electrons of the atoms. In nearly all semiconductors, there is a sizable gap, of order 1 eV, between these two bands, but, in graphene, this gap is zero. Since the gap is zero, the two bands have a linear dispersion with momentum as

$$E = \pm \hbar v_F \, |k| \, , \tag{9.5}$$

where the positive energy states are the conduction band and the negative energy states are the valence band. These bands are often called Dirac bands in comparison with his relativistic version of quantum mechanics, discussed in Chapter 5.

The interesting aspect is that we can measure these bands. The approach used is a variation of the photoelectric effect explained by Einstein in 1905 [24]. Here, high energy photons are directed onto the surface of the material and the electrons that come out are analyzed for their energy and momentum. Because angle resolution is used on both source and emitted electrons one can determine both the momentum and energy the electron had prior to emission. This technique is termed angle-resolved photo-emission spectroscopy (ARPES), and is a common analytical technique in modern condensed matter physics. This approach allows one to map the part of the energy band near the surface of the material (near the surface, as the electrons can only escape in this region). But as graphene is a monolayer, it is therefore nearly all surface. In Fig. 9.3, we show the results of such a measurement on graphene [25]. In panel (a), one clearly sees the dispersion of the valence band energy as a function of momentum. The point near the top where the bands cross is referred to as the Dirac point, and in this case the Fermi energy lies in the valance band due to *p*-type doping. Panel (b) describes the hexagonal Brillouin zone in momentum space and identifies the symmetry points noted in panel (a). In panel (c) is a blowup of the energy bands at the K point. Here, *n*-type doping has been introduced to assure that states in this energy

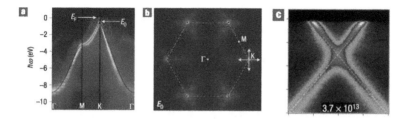

Figure 9.3 (a) The experimental energy distribution of states as a function of momentum along principle directions. (b) The hexagonal Brillouin zone identifying the momentum points identified in (a), along with some constant energy maps (the small circles). (c) Enlargement of the experimental energy bands through the K point for *n*-type doping. Adapted from Bostwick et al. [25] by permission from Macmillan Publishers Ltd.: *Nature Physics,* copyright 2007.

range are occupied by electrons. (The photo-emitted electron has to come from an occupied state.) Now, what is interesting here is that this measurement is being continuously made, yet the quantum structure of the graphene does not collapse.

Indeed, it is important to realize that condensed matter materials are usually subjected to a large variety of measurements that probe the nature of the quantum description of the material. Yet, the quantum system is exceedingly robust and shows no sign of collapse with the measurements. To be sure, the electrons that are emitted have become classical in their description. In a sense, this set of classical electrons is the symbol of Bohm's assertion that the measurement merely picks out a part of the system, in this case the emitted electrons. But the quantum system is not a flimsy construction that evaporates any time someone tries to measure it. And this is very important. The wave nature of the electrons within the semiconductor leads to the energy bands, and this led to the invention of the transistor [26]. The transistor, and the subsequent development of integrated circuits, led to the information age in which we still live. One needs no greater proof of the validity of quantum mechanics than the impact which the semiconductor device and microchip have made on our everyday lives. But this is a quantum mechanics that has greatly evolved from that of the early pioneers and their focus on the hydrogen atom. These quantum systems are extremely robust, even at room temperature and above. They do not just collapse when observed.

Quantum Jumps

As we remarked above, Born was probably considering a method to explain Bohr's ideas of quantum jumps as he worked out the probability ideas in 1926. During this same year, Schrödinger was invited to visit Bohr in Copenhagen, and made the trip in September. As remarked in Chapter 3, Heisenberg was here at the same time. Needless to say, there were long "discussions" on the proper form of quantum mechanics. Bohr attempted his traditional approach to brain washing—sleep deprivation and harangues late at night—primarily to induce Schrödinger to accept the ideas of quantum jumps, a view that was essential to the matrix mechanics and Bohr's philosophy. But Schrödinger would have none of it. In fact, he remarked [27],

> if all this damned quantum jumping were really here to stay, I should be sorry I ever got involved with quantum theory"

But the idea of the quantum jump has remained part of the consciousness of quantum mechanics primarily because of the continued belief in Bohr and his philosophy which emphasized reality existing only with measurement.

The question that we may ask is "are there really quantum jumps?" Well, the issue is both one of time and one of philosophy, at least within the interpretation that you wish to apply to your quantum mechanics. As pointed out, within Heisenberg's matrix mechanics, a particle disappears from one state and reappears in another state. Bohr would enhance this by pointing out that the electron must sit in one of the quantum states, it cannot exist in the energies between the states. Hence, it must jump from one state to the other if a photon is emitted or absorbed. Now, Schrödinger had no need for such jumps as his version of quantum mechanics was based upon the wave picture. In his development of perturbation theory, it could be seen that the wave function for one state would evolve into the wave function for the other state quite naturally under the action of the perturbation [28]. However, the wave was not a particle, and one had to ask where the particles fit into this picture. Both men were speaking from different sides of the problem, and, at least in Copenhagen, they would never come to agreement. Any such agreement would require one of the men to bless the work of the other and to accept their restrictive view. Of course, the issue was discussed in 1927 at the Solvay conference, but without any resolution, and it remained Schrödinger, de Broglie, and Einstein versus Bohr, Born, and Heisenberg.

Then, Geltman asked the question [29], "How long does it take to ionize an atom?" Such a question jumps right to the heart of the matter. He wanted to know the time evolution of the electron within the atom as it was excited out of the atom by a pulse of coherent radiation; e.g., by a single frequency photon. Did the electron instantly leap out of the atom while absorbing the photon? Or, was there a gradual transition over time? These two choices are of course those of Bohr–Heisenberg and of Schrödinger. As he used the wave function approach, he was able to show that the transition gradually occurred over a time scale of a few oscillations of the electromagnetic wave representing the photon. Here, both the wave and the particle were involved, as the wave created the perturbation, but the photon set the amount of energy that was exchanged. A similar determination was done by Lane [30] in analyzing the decay of a state created by a local electric field. Lane pointed out that the approach to the completed transition was characterized by an oscillatory behavior, which could actually make the probability on the short time scale larger than the final value. This was interesting, and these were the early quantum mechanical approaches to answering an old question.

Classically, the treatment of a collision duration may be tracked back at least as far as Lord Rayleigh [31]. Quantum mechanically, however, the problem is better raised as asking just how long it takes to establish the perturbation result, often referred to as the Fermi golden rule, by which the quantum transition is determined. For scattering of an electron by a phonon, this was first discussed by Paige in studies of the transport in Ge and by Barker in high electric field quantum transport [32, 33]. However, these latter approaches used an estimate based upon the energy dependence of the matrix elements rather than trying to analyze the time dependence within the standard formulation of first-order time-dependent perturbation theory. This latter seems to have been first done by Geltman [29], as mentioned. To understand how the process works, we need to turn to the time scale involved.

If you ever watch one of the early movies from Hollywood, the frames tend to jump from one to the other. This might lead you to presume that the actors are actually making such jumps, but with better technology, the motion in movies became quite smooth. Similarly, if you are observing the quantum transition on the scale of life—seconds, minutes, hours, etc., the motion might just look like jumps. However, if you look on the truly short time scale, say femtoseconds, then the transition is not a single jump. What we now know happens when a photon, with its electromagnetic field, approaches a semiconductor, is that the process is far more complicated. Yet, if I am interested only in the fact that solar photons are absorbed in my p–n junction solar cell to produce current, it doesn't matter. On the other hand, if I want to understand the underlying quantum physics of the process, then the correct approach becomes critical. In the idea of ionizing an atom from an incident electromagnetic field (and photon), it is important to understand that the field first creates a correlation between the initial and final state, which can lead to Rabi oscillations between these two states [34]. That is, the wave function for a state in the conduction band becomes entangled with a wave function for the electron in the valence band. These were the oscillations that both Geltman and Lane observed. This correlation is called the polarization, and the first step of the electromagnetic wave of the photon is to create this polarization. The conjugate field then breaks up this correlation, leading to the completed process. This may result in the electron remaining in the valence band and the photon going happily on its way, but it most likely results in the electron moving from the valence band to the conduction band, creating the electron and hole needed in the solar cell. The fraction of photons which actually create the electron–hole pairs is the quantum efficiency and is approaching 100% in modern solar cells. While it is the electromagnetic field that creates the interaction, it is the energy of the photon that is transferred in creating the electron-hole pair. The process is precisely that outlined by Geltman, but applied to the condensed matter system.

The proper quantum mathematics for the entire process was applied later by Haug [35] and by Rossi and Kuhn [36] in the treatment of very fast laser excitation of semiconductors. Hence, the electromagnetic field first creates the correlation (or polarization) between the bands, and then the completed electron–hole pair creation was the second step, just as described above. In this process, both Geltman and Lane [33, 34] found that the approach to the asymptotic form, given by the Fermi golden rule, for long times shows weak oscillations about this asymptotic result, just as mentioned above. These oscillations center on the asymptotic result, indicating that the probability of scattering (or emission) is actually enhanced for short periods of time, although these oscillations die out rather quickly. The entire process takes just a few femtoseconds, and this makes it hard to experimentally observe. However, the above is the mathematical procedure. Experiment is required to confirm that the theoretical approach is correct. Measurement of the fast polarization requires techniques of nonlinear optics, and femtosecond pulsed lasers, and these have arrived only in the last few decades. The important approach is known as time-resolved four-wave mixing, in which multiple coherent light sources are used to probe the process. The incoming laser beam creates a *lattice* of particles in the material. The subsequent laser light is diffracted from this lattice, and the amount of diffraction is determined by a particular component of the polarization. Hence, the time resolved studies can determine the time dependence of the polarization, even on this femtosecond time scale. Indeed, the experimental results can even image the polarization and its time response, clearly establishing the correctness of the approach.

Returning to the electron–phonon interaction discussed above, an approach based upon the above optical procedures has been carried out [37]. In this case, it was not the electromagnetic wave, but the lattice vibrational wave accompanying the phonon which instigated the interaction. Nevertheless, the process is quite similar, in that a polarization is built up between the initial state of the electron and the final state of the electron. Then this polarization decays as the "particle" completes the scattering process. The vibrational wave creates the interaction leading to the scattering but the quantized energy of the phonon determines the amount of energy exchanged in the process, while the wave determines the momentum exchange. Again, the entire scattering process for optical phonons requires just a few femtoseconds—small but sometimes measurable, as this "collision duration" could possibly be seen in femtosecond optical excitation of semiconductors [38]. Thus, it seems that most modern scattering processes which are governed by perturbation theory occur in this manner.

Modern approaches to time-resolved studies require these fast lasers simply because they provide the finest time scale over which measurements can be made at the present time. The important point here is that they give us a view into the basic operations of quantum transitions. On these time scales, one can actually observe the time response of the perturbation interaction. And, on these scales, Schrödinger was right—there is no need to demand that quantum jumps are the only correct way to think about the quantum transition.

Weak Measurements

One of the arguments of EPR was for the existence of a reality in the system before the measurement. Can this be made to occur even when others would insist upon the collapse of the wave function. In recent years, there have been new ideas on this thought, which perhaps can best be described by weak measurements [39]. In such an experiment, for example a measurement of the spin of an electron, a value is found which is far from the "allowed" values of just "up" or "down", $+\frac{1}{2}$ or $-\frac{1}{2}$. The authors refer to this as a "... *weak value of a quantum variable.*" As we discussed in the section above on the collapse of the wave function, measurement is supposed to project the system into one of the eigenstates (either of the system or of the measurement apparatus, depending upon who you ask). This collapse is supposed to be one of the fundamental tenets of quantum theory. And, as we discussed in Chapter 6, the eigenstates for the spin are supposed to be just the two values mentioned. For this approach, these authors suggest [39]:

> We start with a large ensemble of particles prepared in the same initial state. Every particle interacts with a *separate* measuring device, and then the measurement which selects the final state is performed. Finally, we take into account only the "readings" of the measuring devices corresponding to the post-selected particles.

By using this scheme, for example in the Stern–Gerlach experiment, they are able to obtain values other than the eigenvalues mentioned above. In this approach, the quantum system is only weakly coupled to the measuring device, just enough to indicate a measurement, but it is assumed that this does not disturb the quantum system to any great extent. This seems to contradict some of the basic tenets of quantum theory, at least as they arise from Bohr's ideas. Yet, it is argued that these weak measurements lie within the boundaries of quantum theory and do not violate any fundamental concepts. This is easily supported if one takes the view of reality that the system has a well-defined wave function, even if it is spread out in its representation of a particle, even before the measurement is made.

Of course, there are problems with the ideas, in that such measurements might be thought to be incompatible with the standard Copenhagen–Göttingen viewpoint. This viewpoint might well be expressed in the modern world as has been done by Busch [40]:

> the inevitability of disturbance and entanglement in a measurement, the impossibility of repeatable measurements for continuous quantities, and the incompatibility between conservation laws and the notion of repeatable sharp measurements.

While the earlier measurements were simple spin measurements, more modern approaches have concentrated on a continuous measurement in which one integrates the weak signal over time [41]. Hence, one gradually learns more about the system [42].

Here, again, the continuous measurement of the quantum system is essentially what was being done in a previous section with the ARPES measurements of graphene, although blasting a monolayer of graphene with relatively energetic photons would not be my idea of a *weak* measurement. One problematic aspect of continuous measurements is the time evolution of the system, where what is being measured may be a function of time. Such a problem is likely to arise in a system which lacks time reversal symmetry, due for example to a dissipative problem, but this has also been addressed with weak measurement theory [43].

The point here is that modern physicists have managed to find ways to make measurements on a quantum system without collapsing the system, or its wave function. Certainly, some disturbance arises, but does not wipe out the system. Moreover, from these types of measurements, one has to ask whether the projection postulate put forward by Dirac and von Neumann is the full story of measurements in quantum systems, or is there more to be understood. But it is also quite clear that, for a system to undergo continuous measurements and remain a quantum system, there must be a degree of physical reality connected to the system, and this reality is quite likely to exist before the first measurements.

Decoherence Theory

In the previous chapter, the consistent histories approach to quantum mechanics was discussed. This approach is often called the GGHO approach after Griffiths, Gell-Mann, Hartle, and Omnès [44–46]. Let us review this method here. In the usual quantum mechanics, wave function collapse is considered the byproduct of measurements. In contrast, wave function collapse in the consistent histories approach is a mathematical procedure for calculating conditional probabilities during the quantum evolution [47, 48]. Probabilities are introduced as part of the axiomatic foundations of quantum theory, irrespective of measurements. We can think of a history as a sequence of alternatives at a specific set of times [49, 50]. These alternatives may be simple yes/no statements, but open the door for probabilities at each state. Thus, the states $|\psi\rangle$ at any point in time may each be characterized by a "probability" P_k for the possible states at that time. In the strictest interpretation, this probability is a projection operator which takes the entire Hilbert space to the allowed set of states. In this sense, the system is projected in the sense of Dirac [19], as mentioned above. The succession of the various states through which the system moves creates a trajectory in the spirit of Dirac [51] or Bohm [52]. In moving from one state to the next, it may be assumed that there is a time ordering in the states, with $\ldots t_k < t_{k+1} < t_{k+2} \ldots$, etc. By nature, the projection operators, or probabilities, at each time step compress the phase space and introduce decoherence, which is why the consistent histories approach is often called a decoherence approach.

While the GGHO approach uses decoherence as a tool to generate the trajectories, it is not the only suggestion for an interpretation of quantum mechanics which relies upon the concept of decoherence. Coupled to the ideas of measurement in quantum mechanics and the survival of the quantum nature is the question of the connection between quantum and classical systems. Zurek [53] has investigated exactly this aspect through what is also often called a decoherence theory, and we will see how decoherence is crucial to this approach. We are particularly interested in this theory, because it has a natural parallel in experiments which have been made in semiconductor heterostructures. Between detailed measurements, and quantum simulations of the experimental structures, many of the observations are very consistent with this latter decoherence theory. And, consistent with the above measurements of, e.g., ARPES, the quantum mechanics of the condensed matter system does not vanish with the measurements.

To begin, decoherence has been thought to be an important part of the measurement process, especially in selecting the classical results; that is, in passing from the quantum states to the measured classical states of a system. However, the description (and interpretation) of the decoherence process has varied widely, but an important part of this is the interaction of the system upon the environment, as well as the interaction of the environment upon the system. Let us stop here for a moment and explain this point. A closed quantum system is fully coherent because it is time reversible as mentioned earlier. It satisfies the Schrödinger equation, which itself is time reversible. If we want to make a measurement, we have to open the system. The measuring system exists *only* in the external environment that the quantum system will see when it is opened to the outside world. Yet, if this apparatus is to measure some property of the quantum system, it must interact with it. Zurek has proposed that the interaction on the system by the environment leads to a preferred, *discrete* set of quantum states, known as *pointer* states, which remain robust. These are the states in the quantum system that do *not* interact with the environment. Other states, due to their interaction with the environment, have their coherence reduced by the decoherence process [53]. That is, the quantum properties of what was the closed system are not all destroyed by the interaction with the environment. Rather, the formerly closed system still retains a set of quantum states that maintain their quantum nature. Hence, when the system is opened, the total set of states is split into two groups. One group strongly interacts with the environment, and is subsequently decohered— e.g., it loses its coherent quantum properties. The second group does not interact effectively with the environment and remains coherent. In Zurek's view, the decoherence process, which works on the first group of states, effectively defines the second group of states. His decoherence-induced selection of the preferred (second) group of states, which he calls the *pointer* states, is termed *einselection* [54].

The einselected states, those states within the formerly closed quantum system and that still retain their quantum mechanical character, have a number of important properties. First, and foremost, is that they are extremely stable, in that they have withstood the interaction with the external environment. In some sense, this is fully in line with Bohm's view [52], as that reduced part of the system which is measured is actually the decohered part of the system which interacts with the environment. In addition, however, Zurek [53] points out that they are closely related to the trapped *classical* orbits that would exist in the same structure. Because of this, the statistical nature of these states should possess classical statistics. This is an interesting prediction, but it is one that can be checked with both simulation and with experiment, so in the remainder of this section we discuss just these points.

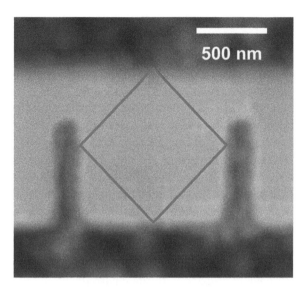

Figure 9.4 Atomic force microscope image of the quantum structure. The dark areas have been etched away, leaving the heterostructure. The dot is defined by the top and bottom edges and the two vertical trenches that have been etched into the structure. The red diamond is discussed in the text, but is not part of the structure.

One possible system in which to study this view of quantum measurements, and sustainability of the quantum state, is an open quantum dot. The dots that I discuss here are created in the interface between layers of two different semiconductors, known as a hetero-interface. At the interface between these two materials, a quantum well forms and this is populated with electrons whose motion perpendicular to the interface is quantized. However, they are free to move along the interface between the two materials. We now limit this lateral motion by creating potential walls which confine the electrons into a small area, typically of micron size. Such a structure is shown in a top down view in Fig. 9.4. In the figure, we show an atomic force microscope image of the overall device (the goldish colored region). The electrons are confined in a thin layer of InAs, which is sandwiched between two wider bandgap materials. The structure has been etched into the form shown in the figure, and the two vertical trenches, as well as the horizontal edges provide the confinement walls. The dot is shown open to the left and right regions which are part of the environment. In this case, the environment is the two-dimensional electron density outside the dot, and contacts are made to this region relatively far from the dot. Other structures are formed from GaAs and AlGaAs, and then defined by metallic gates placed upon the top surface. But the active dot region is shaped quite similarly to that shown in the figure.

Experimental studies of the dots are carried out with what is called conductance spectroscopy. The dots are studied at very low temperatures, where the quantum effects are observable. A small magnetic field is applied and oriented normal to the surface shown in the figure. Then the conductance is measured as the voltage applied to the confinement gates, or to the back gate in the device in the figure, and the magnetic field is varied. Obviously, a reasonable current passes through the device, and is carried by the decohered states which have hybridized with the environment states. On top of this signal, however, is a nearly periodic oscillation, or current fluctuation, which is attributed to the presence of the einselected pointer states [55, 56]. The question is just how does this connection occur, since the pointer states do not connect to the environment. The answer is through phase space tunneling [57], which can only be described as a very weak measurement process. In this process, an electron can enter via the current carrying states and then tunnel onto and off of the pointer state. When this happens, a peak occurs in the conductance through the system. Along with the experiments, quantum simulations of the transport through the dot show precisely the same behavior as that observed in the experiments [58]. In these simulations, a particularly dominant pointer state has been found to exist, and this state has the generic form of the red diamond depicted in Fig. 9.4.

If we examine the red diamond in the figure, we note that the four corners are located at the edges of the active dot. At these edges, the wave function must vanish. Hence, just as described by de Broglie [59], we have to fit a multiple of a half wave-length into each of the four arms of the diamond. A typical quantum dot, with the dimensions indicated in Fig. 9.4 (about 1 micron square), will have about 4000 electrons in it. Thus, a great many quantum states are filled. As we fill these states, multiple copies of the red diamond, each with a different number of half wavelengths in each arm all lie on top of one another. These reinforce each other and make the diamond pattern observed in the simulation extremely stable. The current entering (and leaving) the dot passes through the two constrictions at the top of the structure. These are quantum structures in their own right, in which a half wavelength has to fit through the vertical opening, and are called quantum point contacts. In general a great many modes pass through these contacts, but each mode shows the presence of the diamond pointer state [60]. Zurek remarks on the redundancy of these indicators of the pointer states in the environmental coupling as "their 'fitness' in the Darwinian sense" [53]. He goes on to express the stability as quantum Darwinism [61]. The fact that a great many states have this same general structure leads to the periodic fluctuations in either gate voltage or magnetic field, as each fluctuation occurs as one of the states passes through the Fermi energy. Hence, the observation of the fluctuations is a sign that the experiment is continuously measuring the quantum properties of this system, and that the quantum system remains very robust.

We remarked above that the fluctuations are very periodic in gate voltage or in magnetic field [58]. We can understand this from the above argument on fitting multiples of half wavelengths into the side of the red diamond. This means that we need to have the wavelength satisfy $\lambda = 2a/n$, where n is an integer. This leads to the wave vector being described as $k = 2\pi/\lambda = \pi n/a$. As we vary the gate voltage or the magnetic field, we move these levels through the Fermi energy, which is the energy at the top of the electron distribution and is set by the electron density in the environment. A tunneling peak will be observed each time one of the states moves through this level and that corresponds to $\Delta k = \pi/a$, or a change in n by a single unit. The energy varies as the square of k, or $\Delta E \sim k\Delta k$. Hence, the period of the oscillation should vary as a, or as the square root of the area of the dot ($A = 2a^2$). In fact, this has been observed both experimentally [55, 62] and in direct simulation of the dots [63]. This holds as long as the dot is not too small. In very small dots, k also directly starts to show its dependence on the edge, and the period begins to vary with the area of the dot, a fact also found in simulation and experiment [63].

It was also remarked that Zurek's decoherence theory says that the ensemble of pointer states should have classical statistics [53]. There is two parts to this. First, the behavior of the pointer states should be quite similar to the equivalent classical regular orbit in the same cavity. Then, the population of the pointer states as a function of energy should have a classical distribution. To check the first point, the self-consistent potential of a GaAs quantum dot was generated from detailed characterization of the experimental device. Once the potential was known, then classical simulations could be carried out, and it would be found that these classical results compared very well with the quantum simulations, both for the form and the periodicity [57, 64]. Then, the statistics were computed for the ensemble of pointer states following a technique described for chaotic systems [65]. For quantum systems, one would expect to have one of the Gaussian statistical ensembles. However, it was found that the pointer states satisfied the classical Poissonian distribution [66], as expected by Zurek. In addition, it was demonstrated that there was no way to fit the data to one of the Gaussian ensembles. So, it was clear that the pointer states obeyed classical statistics.

Finally, it is of interest to see if one can actually observe the presence of the pointer state. Since the pointer state is a strong wave function, there will be a high probability of an increased electron density corresponding to the probability given by this wave function. We can study this density with a technique called scanning gate microscopy (SGM) which utilizes the same atomic force microscope with which the dot was imaged in Fig. 9.4. In SGM, the scanning tip, which is normally a dielectric, is metallized so that a voltage can be applied to the tip. Then, as this tip is scanned over the device, this voltage perturbs the electron density, and a change in conductance can be seen. The difficulty of course is that the pointer states have appeared because they do not sense the environment. Now, we are going to bring the environment, in the guise of the SGM voltage, into the system where the pointer state lives. This has all the signs of a destructive measurement, even classically. A classical electron would follow an orbit which just deflects around the SGM tip and would not appear strongly in the measurement. With the quantum system, however, the constraint on the wave-

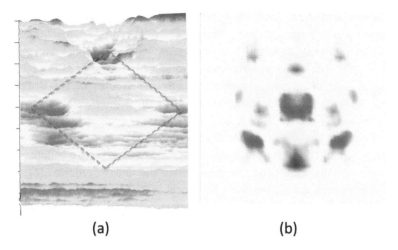

(a)   **(b)**

Figure 9.5 (a) An SGM image of what is believed to be the dominant pointer state in the dot of Fig. 9.4. (b) Results of a quantum simulation of the SGM process in the dot.

length limits this deflection, and hence the pointer state is likely to be affected much more. The best we can hope for is to see the region where the pointer state existed before the voltage appeared. And this works out fairly well. In Fig. 9.5a, we show an image of the change in conductance as the tip is rastered over the surface of the dot [67, 68] (the red regions are where the electron density was higher presumably due to the pointer state). A dashed blue line is also shown as a guide to the eye of where the pointer state was expected to be (the image is actually tilted away from the viewer to better image the contrast). In spite of the objection to measurement on the part of the pointer state, the image bears a reasonable similarity to the red diamond in Fig. 9.4. For comparison, a simulation of the quantum system with the scanning gate potential is shown in Fig. 9.5b, and it can be seen that there is still a representation of the pointer state, but that it is not very distinct, which agrees with the experiment. In fact, the robustness of the pointer state during the measurement is taken to be support for the ideas of quantum Darwinism [61].

Now, it is fairly clear from the discussion of the above experiments and simulations that at least this one typical quantum system satisfies almost perfectly the expectations for Zurek's decoherence theory. The pointer state ideas, the connection to classical dynamics, and the stability expected from quantum Darwinism, all seem to be observed in these experiments. And it is important to note that this is not just a single experiment, as the studies of the dots and their behavior in applied voltages and magnetic fields is a multi-decade endeavor [69, 70]. The evidence is overwhelmingly in favor of the decoherence theory. The quantum systems are robust under continuous measurements and do not seem to collapse when initially measured. In this sense, the measurement ideas of Bohm and Zurek seem to be more in keeping with what is observed in experiments. It is also important to note that the close connection between classical behavior and quantum behavior is one fully in tune with many of the pioneers of quantum mechanics. Moreover, the quantum system seems to be stable under continuous measurement, and also seems to be consistent with a state of reality prior to any measurement being made.

Waves and Particles Again

The conundrum of quantum mechanics was presented quite early by the two-slit experiment. The wave concept was represented by the interference fringes discovered by Young, while the particle concept was represented by the individual photon strikes observed on the measuring screen. This was reinforced by the measurements of Taylor in 1909 [71]. And at the same time, Einstein showed that both the particle and wave approach were necessary and, in fact, could be observed as distinct contributions to the fluctuation spectrum [72]. In modern times, Taylor's experiment has been reproduced using entangled photon pairs, generated by spontaneous parametric down conversion. In this latter process, an energetic photon passes through a nonlinear medium, and is converted to two lower energy photons, in this case at 842 nm and 776 nm wave-lengths [73]. This allowed a time resolved study of the interference and the phase relationships. Nevertheless, the assertions about not being able to simultaneously measure both the interference and determine which slit the photon went through remained as strong as ever. It was rephrased as "duality" in which "Simultaneous observation of wave and particle behavior is prohibited, usually by the position–momentum uncertainty relation" [74]. At least with this view, the magic of quantum mechanics, and the philosophy of Bohr, remained as strong as ever, but the onset of weak measurements has challenged this prohibitive view. A series of experiments in the past decade appear to provide a serious counterpoint to the so-called duality claims.

Another manner in which to present the duality principle is that complementarity tells us that if information is available on one complementary variable, then even if we chose not to measure it, we will lose the information on the other complementary variable. This was, of course, at the heart of the Einstein–Podolsky–Rosen paradox, and is at the heart of the two-slit experiment. Yet, Menzel et al. [75] have used the parametric down-conversion process to generate entangled photon pairs, and to obtain which-slit information on the signal photon by a coincidence measurement of the two photons. Yet, they observe in a different coincidence measurement interference fringes in the signal photon. This surprising result is contrary to the principle of complementarity. The authors have suggested that this result is possible because of the transverse mode structure which produces a superposition of two intensity maximum representing two macroscopically distinguishable wave vectors for the signal photon. Hence, it appears to be not exactly a single photon, but a superposition.

Perhaps a more relevant experiment was carried out by Kocsis et al. [76]. In this experiment, single photons generated from a semiconductor quantum dot laser are passed through a double slit apparatus, which of course creates the interference fringes, but through a set of weak measurements of the photon momentum, which was post-selected according to the results of strong measurements of photon position as they passed a series of planes. In the experiment, the quantum dot is an InGaAs dot embedded within a GaAs/AlGaAs heterostructure, which itself is patterned into a vertical optical cavity. The dot is optically pumped and emits single photons at 943 nm wave-length. The single photon nature is confirmed by correlation measurements. These photons are then coupled into a single mode optical fiber and sent through a 50:50 beam splitter into one of two paths. The fibers are twisted around so that the two photon paths exit the fibers and are then reflected into two parallel paths by mirrors. This produces two Gaussian beams which are spatially separated by about half a centimeter. The weak measurement is achieved by a piece of birefringent material (propagation velocity is different in different polarization directions), calcite in this case. The angle of incidence of the photons upon this crystal depends upon their transverse momentum. The birefringence induces a phase shift proportional to the transverse momentum of the photon. The linear polarization is thus converted into an elliptical polarization which serves as a pointer to the momentum of the photon. This elliptical polarization is then projected into the circular basis by a quarter wave plate, and the photons are then passed through a beam displacer. The beam displacer transmits the right-hand circularly polarized component unchanged, but displaces the left-hand circularly polarized component vertically by about 2 mm. When the photon impinges upon the CCD detector array, these two components can be separated out and the momentum determined. But the ensemble of photons creates the interference fringes in the detector array, as a strong measurement. Hence, they are able to obtain a measure of both the photon position and momentum, and these are averaged over a great many photons. By thus reconstructing the photon trajectories, they find that these photons seem to follow typical Bohmian trajectories, as discussed in Chapter 7 and Fig. 7.2.

It has been suggested that such post selection sampling can provide an apparent violation of the duality principle [77]. However, the arguments go back significantly further. In such a "which way" experiment, it was suggested much earlier that the particle's Bohmian trajectories could exhibit contradictory behavior [78]. That is, the Bohm trajectory might suggest one slit when the particle actually went through the other. This, of course, led to some debate among physicists. Finally, however, it was pointed out by Hiley and Callaghan [79] that, if the correct Bohm approach is used, the result is exactly the one expected by quantum mechanics. This then suggests that the experiments of Kocsis et al. [76] have a basis for reality. Subsequently, this latter group used entangled photons to provide a much better mapping of the Bohmian trajectories through the two slits [80]. In this second experiment, the system was much like the first, except that polarization entangled photons are created from a laser. One of each pair is sent through the apparatus as described in the previous paragraph. In this case, if the polarization is horizontal, the photon is prepared in the upper slit wave function, while if the polarization is vertical, it is prepared in the lower slit wave function. The two slits are now only 2.6 mm apart. The polarization of both photons after the slits is made the same by a set of half-wave plates and a polarizer. But the polarization of this photon is still entangled with that of the second photon, and the polarization of this second photon is measured. The momentum of the first photon and the interference is measured as before. The entanglement provides additional information in confirming the position and momentum of the ensemble of photons, and only the trajectories for a particular polarization of photon 2 are used in the reconstruction. The results basically confirm the earlier measurements. As remarked above, the trajectories obtained bear a striking resemblance to Fig. 7.2 in Chapter 7. They conclude that an uncertainty does result, in that Bohmian trajectories arising from passage through the lower slit may actually be associated with "which way" measurements that associated with either slit; e.g., the interpretation may be contradictory. Nevertheless, these experiments provide strong support for the validity of Bohm's view of quantum mechanics.

In spite of whether or not there is some confusion, or ambiguity, about which slit the particle actually passed through, these experiments confirm that the photons are following individual trajectories which are affected by the two slits and the resulting quantum potential. They thus have particle properties, but when the ensemble of photons arrive at the final observation screen, they build up the interference pattern, in complete agreement with Taylor's experiments more than a century earlier. As a result, it is hard not to draw the conclusion that the Bohmian dynamics, and the corresponding interpretation of the physics, are a valid description of the effects. From these sets of experiments, both the early ones and the more recent ones, as well as the statistical results obtained by Einstein on the noise sources, we once more are encouraged to conclude that it is particle *and* wave, not particle *or* wave!

References

1. R. Feynman, *The Character of Physical Law* (MIT Press, Cambridge, 1965), pp. 147–148.
2. E. H. Hall, *Am. J. Math.* **2**, 287 (1879).
3. K. von Klitzing, G. Dorda, and M. Pepper, *Phys. Rev. Lett.* **45**, 494 (1980).
4. B. N. Taylor, *J. Res. Natl. Inst. Stand. Technol.* **116**, 797 (2011).
5. C.-K. Chiu, J. C. Y. Teo, A. P. Schnyder, and S. Ryu, *Rev. Mod. Phys.* **88**, 035005 (2016).
6. M. Born, *Z. Phys.* **37**, 863 (1926); trans. and reprinted in J. A. Wheeler and W. H. Zurek, *Quantum Theory and Measurement* (Princeton Univ. Press, Princeton, 1983), pp. 52–55.
7. M. Beller, *Stud. Hist. Phil. Sci.* **21**, 563 (1990).
8. Y. Aharonov and D. Bohm, *Phys. Rev.* **115**, 485 (1959).
9. M. V. Berry, *Proc. R. Soc. London, A* **392**, 45 (1984).
10. R. Jackiw and S.-Y. Pi, *Phys. Rev. D* **42**, 3500 (1990).
11. B. L. Johnson and G. Kirczenow, *Rep. Prog. Phys.* **60**, 889 (1997).
12. K. Hess and W. Philipp, *Proc. Natl. Acad. Sci.* **98**, 14224 (2001); **98**, 14228 (2001).
13. D. Smirnov, H. Schmidt, and R. J. Haug, *Appl. Phys. Lett.* **100**, 203114 (2012).
14. C. Davisson and C. H. Kunsman, *Science* **54**, 522 (1921).
15. G. P. Thomson and A. Reid, *Nature* **119**, 890 (1927).
16. W. Heisenberg, *Z. Phys.* **33**, 879 (1925); trans. in B. L. van der Waerden, *Sources of Quantum Mechanics* (Dover, Mineola, New York, 1967), pp. 261–276.
17. W. Heisenberg, *Z. Phys.* **43**, 172 (1927); trans. in J. A. Wheeler and W. H. Zurek, *Quantum Theory and Measurement* (Princeton Univ. Press, Princeton, 1983), pp. 62–86.
18. G. Bacciagaluppi and A. Valentini, *Quantum Theory at the Crossroads* (Cambridge Univ. Press, Cambridge, 2009).
19. P. A. M. Dirac, *The Principles of Quantum Mechanics*, 4th ed. (revised) (Clarendon Press, Oxford, 1958), p. 36.

20. J. von Neumann, *Mathematical Foundations of Quantum Mechanics*, Trans. by R. T. Beyer (Princeton Univ. Press, Princeton, 1955).

21. A. Whitaker, *Einstein, Bohr, and the Quantum Dilemma* (Cambridge Univ. Press, Cambridge, 1996), p. 196.

22. J. T. Cushing, *Quantum Mechanics, Historical Contingency and the Copenhagen Hegemony* (Univ. Chicago Press, Chicago, 1994), p. 68.

23. P. R. Wallace, *Phys. Rev.* **71**, 622 (1947).

24. A. Einstein, *Ann. Phys.* **322**, 132 (1905).

25. A. Bostwick, T. Ohta, T. Seyler, K. Horn, and E. Rotenberg, *Nat. Phys.* **3**, 36 (2007).

26. W. Shockley, J. Bardeen, and W. H. Brattain, *Science* **108**, 678 (1948).

27. J. Gribbon, *Erwin Schrödinger and the Quantum Revolution* (Wiley, New York, 2013).

28. E. Schrödinger, *Ann. Phys.* **80**, 437 (1926).

29. S. Geltman, *J. Phys. B* **10**, 831 (1977).

30. A. M. Lane, *Phys. Lett.* **99A**, 359 (1983).

31. Lord Rayleigh, *Phil. Mag.* **11**, 283 (1906).

32. E. G. S. Paige, *The Electrical Conductivity of Germanium*, Vol. 8 of *Progress in Semiconductors*, ed. by A. F. Gibson and R. E. Burgess (Wiley, New York, 1964).

33. J. R. Barker, *J. Phys. C* **6**, 2663 (1973).

34. S. L. Haan and S. Geltman, *J. Phys. B* **15**, 1229 (1982).

35. H. Haug, in *Optical Nonlinearities and Instabilities in Semiconductors*, ed. by H. Haug (Academic, San Diego, 1988); *Phys. Stat. Sol. B* **173**, 139 (1992).

36. F. Rossi and T. Kuhn, *Phys. Rev. Lett.* **69**, 977 (1992); *Phys. Rev. B* **46**, 7496 (1992).

37. P. Bordone, D. Vasileska, and D. K. Ferry, *Phys. Rev. B* **53**, 3846 (1996).

38. D. K. Ferry, A. M. Kriman, H. Hida, and S. Yamaguchi, *Phys. Rev. Lett.* **67**, 633 (1991).

39. Y. Aharonov, D. Z. Albert, and L. Vaidman, *Phys. Rev. Lett.* **60**, 1351 (1988).

40. P. Busch, in *Quantum Reality, Relativistic Causality, and Closing the Epistemic Circle: An International Conference in Honor of Abner Shimony* (Springer-Netherlands, Amsterdam, 2009), pp. 229–256. Also available as arxiv:0706.3526.

41. V. B. Braginsky and F. Y. Khalili, *Quantum Measurements* (Cambridge Univ. Press, Cambridge, 1992).

42. A. A. Clark, M. H. Devoret, S. M. Girvin, F. Marquardt, and R. J. Schoelkopf, *Rev. Mod. Phys.* **82**, 1155 (2010).

43. A. Bednorz, K. Franke, and W. Belzig, *New J. Phys.* **15**, 023043 (2013).

44. R. B. Griffiths, *J. Stat. Phys.* **35**, 219 (1984).

45. M. Gell-Mann and J. B. Hartle, in *Complexity, Entropy and the Physics of Information*, ed. by W. H. Zurek (Addison-Wesley, Redwood City, 1990).

46. R. Omnès, *J. Stat. Phys.* **53**, 893 (1988).

47. R. B. Griffiths, *Consistent Quantum Theory* (Cambridge Univ. Press, Cambridge, 2003).

48. M. R. Sakardei, *Can. J. Phys.* **82**, 1 (2004).

49. J. B. Hartle, *Phys. Scr. T* **76**, 67 (1998).

50. R. Omnès, *The Interpretation of Quantum Mechanics* (Princeton Univ. Press, Princeton, 1996).

51. P. A. M. Dirac, *Rev. Mod. Phys.* **17**, 195 (1945).

52. D. Bohm, *Phys. Rev.* **85**, 166, 180 (1952).

53. W. H. Zurek, *Rev. Mod. Phys.* **75**, 715 (2003).

54. W. H. Zurek *Phys. Rev. D* **26**, 1862 (1982).

55. See, e.g., J. P. Bird, K. Ishibashi, Y. Aoyagi, T. Sugano, R. Akis, D. K. Ferry, D. P. Pivin, Jr., K. M. Connolly, R. P. Taylor, R. Newbury, D. M. Olatoona, A. Micolich, R. Wirtz, Y. Ochiai, and Y. Okubo, *Chaos Sol. Frac.* **8**, 1299 (1997), and references therein.

56. D. K. Ferry, R. Akis, and J. P. Bird, *Phys. Rev. Lett.* **93**, 026803 (2004).

57. A. P. S. de Moura, Y.-C. Lai, R. Akis, J. P. Bird, and D. K. Ferry, *Phys. Rev. Lett.* **88**, 236804 (2002).

58. R. Akis, D. K. Ferry, and J. P. Bird, *Phys. Rev. B* **54**, 17705 (1996).

59. L. de Broglie, *Nature* **112**, 540 (1923).

60. R. Brunner, R. Akis, D. K. Ferry, F. Kuchar, and R. Meisels, *Phys. Rev. Lett.* **101**, 024102 (2008).

61. H. Ollivier, D. Poulin, and W. H. Zurek, *Phys. Rev. Lett.* **93**, 220401 (2004).

62. D. K. Ferry, J. P. Bird, R. Akis, D. P. Pivin, Jr., K. M. Connolly, K. Ishibashi, Y. Aoyagi, T. Sugano, and Y. Ochiai, *Jpn. J. Appl. Phys.* **36**, 3944 (1997).

63. N. Holmberg, R. Akis, D. P. Pivin, Jr., J. P. Bird, and D. K. Ferry, *Semicond. Sci. Technol.* **13**, A21 (1998).

64. R. Brunner, R. Meisels, F. Kuchar, R. Akis, D. K. Ferry, and J. P. Bird, *Phys. Rev. Lett.* **98**, 204101 (2007).

65. M. C. Gutzwiller, *Chaos in Classical and Quantum Systems* (Springer-Verlag, New York, 1990).

66. R. Akis and D. K. Ferry, *Physica E* **34**, 460 (2006).

67. A. M. Burke, R. Akis, T. Day, G. Speyer, D. K. Ferry, and B. R. Bennett, *J. Phys. Condens. Matter* **21**, 212201 (2009).

68. A. M. Burke, R. Akis, T. E. Day, G. Speyer, D. K. Ferry, and B. R. Bennett, *Phys. Rev. Lett.* **104**, 176801 (2010).

69. J. P. Bird, R. Akis, D. K. Ferry, A. P. S. de Moura, Y.-C. Lai, and K. M. Indlekofer, *Rep. Prog. Phys.* **66**, 583 (2003).

70. D. K. Ferry, A. M. Burke, R. Akis, R. Brunner, T. E. Day, R. Meisels, F. Kuchar, J. P. Bird, and B. R. Bennett, *Semicond. Sci. Technol.* **26**, 043001 (2011).

71. G. I. Taylor, *Proc. Cambridge Philos. Soc.* **15**, 114 (1909).

72. A. Einstein, *Phys. Z.* **10**, 817 (1909). An English translation is available at https://en.wikisource.org/wiki/The_Development_of_Our_Views_on_the_Composition_and_Essence_of_Radiation

73. P. Kolenderski, C. Scarcella, K. D. Johnson, D. R. Hamel, C. Holloway, L. K. Shalm, S. Tisa, A. Tosi, K. J. Resch, and T. Jennewein, *Sci. Rep.* **4**, 4685 (2014).

74. M. O. Scully, B.-G. Englert, and H. Walther, *Nature* **351**, 111 (1991).

75. R. Menzel, D. Puhlmann, A. Heuer, and W. P. Schleich, *Proc. Natl. Acad. Sci.* **109**, 9314 (2012).

76. S. Kocsis, B. Braverman, S. Ravets, M. J. Stevens, R. P. Mirin, L. Krister Shalm, and A. M. Steinberg, *Science* **332**, 1170 (2011).

77. E. Bolduc, J. Leach, F. M. Miatoo, G. Leuchs, and R. W. Boyd, *Proc. Natl. Acad. Sci.* **111**, 12337 (2014).

78. B.-G. Englert, M. O. Scully, G. Süssman, and H. Walther, *Z. Naturforsch.* **47a**, 1175 (1992).

79. B. J. Hiley and R. E. Callaghan, *Phys. Scr.* **74**, 336 (2006).

80. D. H. Mahler, L. Rozema, K. Fisher, L. Vermeyden, K. J. Resch, H. M. Wiseman, and A. Steinberg, *Sci. Adv.* **2**, e1501466 (2016).

Chapter 10

What Does It All Mean?

Through the previous chapters, we have described the various interpretations, experiments, and philosophies that define the confusion and disagreements in modern quantum theory. On the basis of all of this, Feynman once said, "I think I can safely say that nobody understands quantum mechanics" [1]. Certainly, if one contends that Bohr and the Copenhagen–Göttingen philosophy correctly describes quantum mechanics, then it is clear that Feynman's statement expresses the problems with interpretation. This is because such an interpretation has its own problems with consistency. This may arise because Bohr was never all that definite in his expressions, and was more known for obfuscation than for clarity. But, as we have seen, there are other "interpretations" which yield more clarity and a clearer connection between classical and quantum mechanics. There are even more interpretations that we have not discussed. All alternative interpretations have at their roots some form of disgust at the Copenhagen–Göttingen philosophy, and all have a desire to seek an alternative truth for the understanding of quantum mechanics. Our problem at hand is the question as to just what the truth may well be.

In the late 1980s, a series of television programs appeared as Ethics in America [2]. One program was titled "Truth on Trial." This particular program dealt with legal issues and how the

The Copenhagen Conspiracy
David Ferry
Copyright © 2018 Pan Stanford Publishing Pte. Ltd.
ISBN 978-981-4774-75-8 (Hardcover), 978-1-351-20723-2 (eBook)
www.panstanford.com

so-called truth was obtained in the courtroom. Each lawyer does his best to present the truth as it supports his client. As the late Supreme Court justice Antonin Scalia remarks, each lawyer is working as hard as he can to produce his own view of the truth, or at least that portion of the truth which best supports his client. That is, each lawyer is describing truth, but not the whole truth (only witnesses have to swear to do this). As a result of the adversarial proceedings, it remains for the judge and the jury to decide which version of the truth is the most believable. Now, I am not judge and jury. Nor does science progress in the same fashion as a courtroom, although for sure science does advance by adversarial proceedings. Theories are put forth as a hypothesis, and experiments may be presented that support the hypothesis. It is expected that supporting experiments which either deny or confirm the original ones are expected from other groups. We expect that correct hypotheses will withstand the preponderance of evidence, evidence that is acquired by extensive investigation, and may grow to be theories. Nevertheless, obfuscation and proof by intimidation are not unknown in the scientific community. And, I am sorry to say, there is ample evidence of the fabrication of experimental results and the subsequent claims, in many fields of science. Yet, quantum mechanics has been with us for a century, and there should now be ample evidence to decide where the reality lies. To the contrary, the debate and the division among scientists continues. Does it continue because of the lack of understanding, or does it continue because of the refusal to accept the results on the part of a fraction of the community? This will have to be decided by you, the reader. To be sure, I have my own views. But, in this chapter, I will try to summarize the situation at the present time.

When we say that the debate continues, we are talking about the fact that there is not a consensus of opinion that would support any one view, and the principles of the debates between Bohr and Einstein remain as vibrant today as they were at the 1927 Solvay conference. Writing some years ago, Jauch summarized the situation in 1971 [3]:

> although quantum mechanics has become the indispensable basic
> theory for all of microphysics and for much of macrophysics as

well, its interpretation has remained a source of conflict from its inception in the late twenties until today. The epistemological questions associated with the so-called Copenhagen interpretation are so profound and revolutionary that many of the founders of this theory . . . have rejected this interpretation . . . .

There have been numerous expositions of these difficulties, some in the form of paradoxes, . . . others consisting of a more formal analysis of the measuring process or the philosophical or mathematical discussions on hidden variables. The fact that these expositions often take the form of controversies with ideological overtones and that they have been going on for more than forty years and have even lately shown signs of increasing in volume and intensity is perhaps sufficient evidence that quantum mechanics has left us with an enigma.

After an additional more than forty years since Jauch's writing, there has been no reduction in either the volume or the intensity of the discussions. Indeed, these discussions continue to be controversial and carry ideological overtones, that are as dogmatic as ever. If one wanted to define a "mainstream," it would be the followers of Copenhagen–Göttingen who consider themselves as the "enlightened cognoscenti" while others are the heretical infidels. Yet, one must note that there are Nobel laureates on both sides of the debate. In fact, today there are more than just two sides of the debate, as there are many alternative interpretations, although we have discussed only a few of these. The main difficulties have been described by Wigner [4]:

the principal difficulty is not the statistical nature of the regularities postulated by quantum mechanics. The principal difficulty is, rather, that it elevates the measurement, that is the observation of a quantity, to the basic concept of the theory. This introduces two difficulties. First, surely, the observation of the position of an electron is not something primitive, and the observation of most other quantities is even less primitive.

Second, however, . . . it seems dangerous to consider the act of observation, a human act, as the basic one for a theory of inanimate objects.

In making this comment, it is all the more strange as Wigner is generally considered to belong to the Copenhagen–Göttingen camp.

The alternative view is of an intrinsic reality prior to the measurement, as discussed by Einstein, and by most of the founders of the wave theory of quantum mechanics. Perhaps, the rationale is best expressed in comments by Selleri [5]:

> Let me state from the beginning that I accept the realistic philosophy. It seems to me very difficult to take seriously a different philosophical standpoint, although it is certainly logically possible. The point is that practically nobody in the society of which the physicist is a part has any doubt about the actual existence of an objective reality outside the observer. By objective I mean existing independently on its being observed.... If the physicist took a different standpoint he would risk to be a totally alienated person, or more probably, to have a split personality, forced as he would be to believe in reality as a man and to decide that reality does not exist as a researcher.

And yet, the "enlightened cognoscenti" insist that this split personality, or dare I say schizophrenia, is the only acceptable view for those who wish to work with quantum mechanics.

Nevertheless, it is useful to summarize the discussions and debates that have been the focus of the previous chapters. In this way, perhaps we can identify possible causes for the views and actions that have been taken by the different players in the field. Of course, we must begin once again with Bohr.

Copenhagen–Göttingen

It is important to point out that what I have been calling the Copenhagen–Göttingen philosophy is not a single unified point of view. Rather, it is a general series of views that are unified by their dislike of the wave theory of quantum mechanics. The dislike is different in different members of this group of founders, and so it is very hard to find a single statement that can describe the so-called Copenhagen interpretation, not the least because Bohr's views seem to have had a time variation. But the problems began with his original set of papers in 1912. In his first paper, he based his development upon the quantization of light put forward by Planck. Yet, the mathematics was a little dodgy and as a result, the Rydberg formula, seemed to appear magically. The Rydberg formula was developed as a description of the energy spectrum of atoms that had been measured experimentally; certainly, Rydberg claimed no theoretical basis. So, Bohr's work was a step forward. In this first paper, he noticed toward the end that the angular momentum seemed to be quantized, so he used this as the basis for his second derivation of the Rydberg formula. Bohr wanted to move away from Planck's quantization so that his work was not considered to be derivative from that of Planck, and beginning with quantization of the angular momentum allowed this approach. Why was the angular momentum quantized? Bohr had no answer to this. Without a reason, the theory could not be considered to be complete—unless the reason didn't matter. If the philosophical view were taken that everything was determined by the experiment, then truth lay in the fact that his derivation gave the Rydberg formula obtained by experiment. The "why" just wasn't important anymore in this philosophy, as it referred to an underlying reality that wasn't there. Reality came from the experiment. Almost everything Bohr did from this time forward was based upon the defense of this philosophy against all comers. Along the way, he seemingly invented new aspects of the philosophy. There was no need for causality or discussion of the deterministic evolution of the quantum system. All knowledge came from the experiment.

There were other problems with the Bohr papers, not the least arising from the fact that he had not done the full three-dimensional model of the atom. It was left to Sommerfeld to do the atom properly a few years later. Importantly, Sommerfeld's proper derivation of the three-dimensional atomic model introduced the azimuthal (or magnetic) quantum number, and this now allowed one to explain one of the Zeeman effects (the splitting of the spectral lines in a magnetic field). This did not seem to bother Bohr, and it certainly did not affect his philosophical interpretation. It seemed only a minor matter that did not reflect upon his view that true reality appeared only with the measurement. Yet, one has to ask why didn't Sommerfeld share in the Nobel Prize for the atomic structure?

One can only speculate at the politics involved. Perhaps it was because scientists from Germany had won three of the previous four Nobel Prizes in physics, and the committee wanted a good, pacifist Scandinavian for the 1922 prize. We will never really know for sure what transpired. We do know that Sommerfeld was nominated 81 times for the prize, including 4 nominations in 1922.

It is clear that Bohr wanted to move away from any discussion of Planck's quantization in his papers on the atomic problem. Kramers discovered this rather roughly from one of Bohr's diatribes (discussed in Chapter 2) prior to his work on the Bohr–Kramers–Slater paper [6]. With the latter paper, Bohr wanted to use Einstein's statistical view of light transitions to try to do away with strict energy conservation, preferring to see it as a statistical average. Unfortunately, energy conservation was observed in each and every interaction in experiments shortly after the paper appeared. Slater had come to Copenhagen hoping to find a deep understanding of the new quantum physics behind Bohr's work. His realization shortly upon arriving that this deep understanding didn't exist led to his early departure as well. Rather, Bohr moved on to new, and usually unproven, assertions, one being that the electron had to jump between quantum states when a photon was absorbed or emitted. This became essential to him, because the electron just couldn't exist except in one of his quantum energy levels. As we saw in Chapter 3, the quantum jump became a real sticking point between Bohr and Schrödinger, with the latter's wave function formulation of quantum mechanics. Schrödinger didn't buy the need for a jump, because he could work through the perturbation theory for the transition without any jump being involved. Nevertheless, the quantum jump became an essential part of the Copenhagen–Göttingen picture, as it was important to Born in his study of scattering and the introduction of the probabilistic picture for quantum mechanics.

It is now clear that one does not need quantum jumps, which means that Schrödinger's view could be considered to be more correct than Bohr's. We saw that light creates a polarization or intermediate state as the photon is absorbed by an atom or a semiconductor. Importantly, this intermediate state can be experimentally measured via nonlinear optics, as was pointed out in the previous chapter. More to the point, if one provides an oscillating driving field that corresponds to the energy difference between two atomic levels, a cyclic behavior of the particle between the two levels occurs, known as a Rabi oscillation or Rabi flop [7, 8]. Indeed, this entanglement between the two levels has become a key element for one form of quantum information processing [9]. Hence, the control of the polarization entanglement and its use has become important to modern applications. An interesting point is that, after getting his degree at Columbia, Rabi had gone to work with Bohr in 1927, but was essentially chased away from Copenhagen [10]:

> we were essentially thrown out of Copenhagen . . . . The ostensible reason was that Bohr had worked very hard that summer, and was tired and not taking people. . . . So, I went to work with Pauli.

A key point of Bohr's arguments with Einstein, particularly with regard to the Einstein–Podolsky–Rosen type of experiments, was that one could not make multiple measurements on the same particle, or quantum system. In his view, the measurement itself defined the quantum system. Thus, two different measurements meant two different, and perhaps diverse, quantum systems. This viewpoint naturally led to the collapse of the wave function for the quantum system. You could not repeat a measurement because the quantum system no longer existed. Born certainly supported this idea, as the need for a probabilistic description resulted from an ensemble of similar quantum systems in order to make multiple measurements, and the collapse of the wave function certainly found its way into the von Neumann treatise [11], as we discussed in Chapter 5. Thus, collapse of the wave function became a crucial part of the Bohr philosophy, but it was a questionable part of his philosophy, as Shimony has remarked [12]:

> Bohr . . . offers no great reward for renouncing a theory of being other that the assurance that we need not try to explain the reduction of wave packets. He has no means proved that speculations concerning the mechanism of the reduction . . . are doomed to sterility.

Indeed, this lack of understanding from Bohr was expressed only slightly differently by Beller [13]:

> Bohr's complementarity, and his operational definition of "phenomenon," is of little help in solving the central interpretive problem of quantum theory—the measurement problem.

The view that the wave function was fully collapsed was not shared by all of the acolytes. Heisenberg in particular talked of making multiple measurements on a quantum system in his original work, where he mentioned [14],

> The second position determination . . . thus limiting the possibilities for all subsequent determinations.

So, he was separating himself from Bohr's position (we have to remember that many of the young acolytes prided themselves on finding "errors" on the part of Bohr). Nevertheless, we now know that continuous measurements can be made on quantum systems without collapsing the nature of the system. Examples of this were given in the previous chapter. We also know that astronomers use the black body radiation from distant stars passing through more local galaxies as a method to measure the absorption spectrum of the latter, and thus to gain information about the elements present. Clearly, the vast stellar system has not collapsed from these measurements. I suppose we must infer that all of these continuous measurements must be categorized as "weak" measurements, but that seems to be a stretch of the imagination.

The Wave Interpretation

So why did Bohr dislike the wave interpretation so much? We can begin with de Broglie, whose approach to the atom provided a reason for the quantization of the angular momentum. And it was rather a simple reason. If we were to accept the wave and particle duality that had been de rigueur since Planck's original paper, then the electron moving around the nucleus must have certain wavelike properties. The most significant is the wave length. De Broglie pointed out that periodicity around the circumference of the orbit required that only integer numbers of wave lengths could fit into the circumference. This, of course, led to quantized values of the energy and angular momentum. Philosophically, however, this pointed out that Bohr had missed the important points in his atomic theory. There now was a real rationale for the quantization of the angular momentum. The fact that it gave a formula in agreement with the experiments was coincidental; there was a solid base upon which the theory was supported. One did not have to wave one's hands anymore over the reason for the quantization. It arose naturally from the interpretation of the electron as a wave. De Broglie had explained what Bohr could not explain. It appears that Bohr could not accept that a young Frenchman had succeeded where he could not. Moreover, it cast doubt on his philosophical viewpoint that the experiment determined everything, and that therefore he didn't need to have a reason for the quantization—it was required by the experimental results.

It grew worse. With the arrival of the Schrödinger equation, quantum mechanics now had a causal, deterministic equation of motion for the wave function. Perturbation theory, based upon the Schrödinger equation, provided transition details that undermined the assertions of quantum jumps. While these transitions occurred much too fast to be "seen" with the experiments of the time, this was no reason to doubt that such transitions actually existed. But Bohr would exclaim that they could not be measured, so had no place in the understanding of quantum mechanics. This was much like Mach asserting that atoms did not exist because he could not see them.

Heisenberg, for his part, never accepted the wave approach. Perhaps this was because it would provide an underlying theory for the uncertainty relations. As we remarked in an earlier chapter, there was a corresponding classical uncertainty in trying to measure position and momentum simultaneously (we used the traveling train analogy for this). Heisenberg's quantum uncertainty went further and asserted that there was a theoretical minimum uncertainty, which occurred between two noncommuting operators. No matter how careful and accurately the measurements were, they could not exceed the quantum uncertainty. The Schrödinger equation allowed one to actually evaluate this uncertainty in terms of a specific wave function. But was the quantum uncertainty deeper than just the fact that the non-zero wavelength which the particle possessed meant that one could not assign it an exact position at any point in time? Heisenberg certainly thought so, but then where did it appear in the wave theory? One could calculate the uncertainty by evaluating the expectation values of the various noncommuting operators, and this fit perfectly well into the wave theory. But it was not apparent that this result didn't merely arrive because of the wavelength being important. Certainly, this had been an underlying thought in Born's development of the probability distribution that followed from the wave function.

Bohr had an additional problem particularly in relation to the uncertainty principle. In his series of arguments with Einstein, he persisted in showing that Einstein's various suggestions would always show some uncertainty, but he was never able to show that this uncertainty went beyond the classical uncertainty that always was there in a measurement. In fact, it never seemed clear to Bohr that he had to do more. He had to show that it was actually the quantum system that was leading to the value found for the uncertainty, and that the measurement uncertainty was approaching (and limited by) Heisenberg's value. Once more, Bohr could wave his hands and pronounce that he had destroyed Einstein's arguments. While Einstein could smile and know that Bohr had not reached the quantum limit, he must have been terribly disappointed that his colleagues did not recognize Bohr's failure. Thus, Bohr could proclaim that he had defeated Einstein, as the world would grow to believe.

The question of wave function collapse seemed to bother the Copenhagen–Göttingen crowd more than it did de Broglie and Schrödinger, perhaps because the latter were not of the opinion that a measurement completely destroyed the quantum nature of the system under test. But the basic problem of quantum measurement has been puzzled over ever since the foundations of quantum theory [15]. It is not a question of probability, because there are problems with simple applications of the probabilistic interpretation. As Schrödinger put it [16],

The orthodox treatment is always that, by way of certain measurements performed *now* and by way of their resulting predictions of results to be expected of other measurements following thereafter either immediately or at some given time, one gains the best possible probability estimates permitted by nature. Now how does the matter really stand?

If one measures the energy of a Planck oscillator, the probability of finding for it a value between E and E' cannot be possibly other than zero unless between E and E' there lies at least one [energy level].... For any energy interval containing [no eigenvalues] the probability is zero. In plain English: other measurement results are excluded.

This view is very important, because it clearly points out that Schrödinger did not accept the complete collapse of the wave function. The wave function was there, and several measurements could be made upon it. Moreover, it is also clear that the probability interpretation can lead to some strange thinking. For sure, there are a number of well-known quantum systems in which the probability of a measurement outside a certain realm will be zero. An electron confined in a quantum well, even an atom, is one such system, but it raises the question in which we base reality upon the measurement: Does a measurement that yields a zero value mean that the quantum system does not exist? And if the observer sees a zero value, does this destroy the system, and what has he learned?

But Schrödinger's comment raises other issues as well. When we think about probability, we normally couple it to the variation of the wave function with position (or, if we Fourier transform it, with momentum). Hence, the probability is one of finding the particle within a small region around a specific position. If we have a Planck oscillator, what probability do we associate with it? The Planck oscillator, or harmonic oscillator, has a series of equally spaced energy levels. There is a specific wave function—the energy eigenstate wave function—associated with each energy level. The statistics of the oscillator however are usually associated with the Bose–Einstein probability distribution, but this tells us what the likelihood is for the states to be occupied at a given temperature. That is, we say that the Bose–Einstein distribution gives us the probability that n bosons are present at a given temperature, for an oscillator of a given energy spacing. But it doesn't tell us the probability of measuring a given energy. Nor does the energy eigenstate wave function tell us the probability of measuring a given energy. It only tells us the probability of finding the particle in position space for that energy. As a result, it is quite difficult to define any probability function for measuring a particular value of energy. What we tend to measure are the photons that are emitted or absorbed by a transition between a pair of these energy levels, as Einstein discussed in 1917 [17]. What Schrödinger may well be inferring from this comment is that it is almost impossible to conceive how a measurement can be constructed that will *define* this quantum system.

Then there is the question of how the measurement of an energy, as in Schrödinger's example with the Planck oscillator, can be made to coincide with von Neumann's reduction of the wave function or even a collapse of the wave function. That there is a difference is expressed by Bohm [18]:

> a measurement of any variable must always be carried out by means of an interaction of the system ... with a suitable piece of measuring apparatus. The apparatus must be so constructed that any given state of the system of interest will lead to a certain range of states of the apparatus.

So, we do not jump into an eigenstate of the measurement system, unless that state is accessible from the quantum system. In this sense, the view is very different than von Neumann's view. But we should go further to point out that designing a measurement to measure a property of, e.g., the Planck oscillator is very detailed and requires in fact some understanding of just what is trying to be measured.

Contrasting Views

As we have pointed out in the above sections, there is a dramatic difference between those who support Bohr's philosophy and those who hold with a degree of reality prior to any measurement. It is a very fundamental difference that lies at the bottom level of interpretation and understanding. One summary of this has been given by Wigner [19]:

> The principal question which is at the center of our . . . discussion concerns the reason for the statistical, that is probabilistic, nature of the laws of quantum-mechanical theory. The equations of motion of both quantum mechanics and of classical theory are deterministic; why, then are the predictions not uniquely given by the inputs?
>
> As far as I can see, there are three possible reasons for this. The *first*, and perhaps most natural, one is that the input is always incomplete, that even if we have as complete a description of the object on which a measurement is undertaken as quantum mechanics can provide, the same is never true of the instrument with which we carry out the measurement. . . . The probabilities which such an assumption can yield for the possible outcomes of a measurement cannot be made to coincide with those postulated by quantum-mechanical theory.
>
> The *second* possible reason for the probabilistic nature of quantum theory's conclusions concerning the outcomes of measurements is that the theory cannot completely describe the process of measurement, that some part of the process is not subject to the equations of quantum mechanics. This is the view which I believe, most of us, including von Neumann and also myself, accept. . . .
>
> The *third* way to avoid the apparent contradiction between deterministic equations and probabilistic events . . . is not to consider the equations as descriptions of reality (whatever that term means) but only as a means for calculating the probabilities of the outcomes of observations.

It seems clear that Bohr accepts only the third way—there is no reality prior to the measurement. Moreover, we can easily recall his arguments against Einstein's suppositions that there was reality prior to the measurements. Hence, the equations cannot be considered as descriptions of reality in Bohr's world. It is clear that classical theory also is subject to both the first and the second reasons given by Wigner. This is especially true in chaotic systems where a miniscule difference in initial conditions can lead to exponentially diverging trajectories for the solutions to the classical equations. It is also true that we quite often cannot completely describe the process of measurement, and every experimentalist knows full well that the measurements are going to be probabilistic. It can only be hoped that the results are clustered sufficiently well to draw some conclusions from the data. Why then should quantum mechanics be singled out, so that similar results are just not acceptable by the cognoscenti?

The close behavior between quantum mechanics and classical theory should in fact be accepted, if for no other reason than that it is relatively easy to transform the Schrödinger equation into the Fokker–Planck equation [20, 21]—the equation for the classical probability distribution function! As for views that differ from Bohr, I think it is fair to say that the close equivalence between quantum and classical dynamics is maintained in the de Broglie–Schrödinger–Bohm interpretation of quantum mechanics. Other supporters of the wave approach would agree with them. Moreover, they point out that this approach gives the same answers to a given problem as are found from any other interpretation.

If we take Wigner's third way from above, we can see the troubles. We know in many quantum systems, we can introduce small changes in the system that will not be observable by a given measurement. That is, if our measurement system cannot detect the change, are we to assume that the change did not exist? How can we then use the changed equations to describe a probability sufficient accurately for the measurement to be changed, if they do not possess a sufficient degree of reality in themselves? Consider a condensed matter system (such as a metal) to which a small magnetic field is applied. We describe each electron in the system by a wave packet whose size is determined by the de Broglie wave length of that electron. Such a wave packet is often described as a minimum uncertainty wave packet, which means that its position and momentum satisfy the uncertainty principle. The charge of the electron is consequently spread throughout the wave packet. This wave packet is composed of the appropriate Bloch states that can exist in the energy band of the system. In the magnetic field, it is found that the wave packet rotates in real space around its average position, and this motion of the rotating charge leads to a magnetic moment associated with the wave packet [22]. Now, this small magnetic moment of each electron modifies the energy bands ever so slightly through an apparent change in the mass, but this magnetic moment is not measurable in transport experiments [23]. Thus, one must define very special measurements to see these magnetic moments of the electrons. So, the question becomes, "Is this a real effect?" If one does not use the proper measurements, it may be concluded the effect is not a part of reality. Yet, if the proper measurements are made, one would conclude that the effect is part of reality. Thus, the question is whether reality is determined by the measurement, as Bohr insists, or is reality a property of the quantum system, whether or not the measurement is ever done? This is the difference between Wigner's third way and the previous two possibilities.

The difference between Wigner's first way and the second way lies in the question of defining the measurement process. In classical dynamics, one really never asks the question as to how the state is going to be measured. Description of the measurement is left to the experimentalist who may have many different approaches. This is basically Wigner's first way. We can illustrate this task with a classic story about a physics test, which included a question about air pressure. The question was,[a] "You are given a fine barometer; how would you use this to measure the height of the XYZ building?" The student submitted an answer on the test different than that expected by the instructor, and was graded down. The student protested that there were several answers possible for such an ill-posed question. Rather than repeat the story, I will paraphrase it in terms of context, all of which relates to the material for which the instructor was testing. If the course was an introductory course on length, time, and gravity, then one acceptable answer might be to tie a string to the barometer and lower over the edge from the top of the building. By measuring the length of string, one

---

[a]The story appears to originate in an article by Alexander Calandra in *The Saturday Review* in September 1968, but has been elevated to an urban legend in physics.

would immediately know the height of the building (which was the answer the student submitted). On the other hand, one could throw the barometer off the top of the building and measure how long it took for the barometer to hit the ground. By the proper use of Newton's laws of physics, one could then determine the height of the building with the well-known expressions for gravity (and even correct for air resistance). If this were a course in trigonometry, an entirely different approach is warranted. Here, one could place the barometer on the ground, and then measure the length of the building's shadow and the length of the barometer's shadow. Using the properties of equivalent triangles, one could then determine the building's height from the height of the barometer. The same problem could be given in a test in the field of business administration, where one could take the barometer to the building superintendent and open negotiations with the ploy: "Look at this magnificent barometer. I will give you this wonderful device if you will tell me the height of this building." Clearly, all of these answers are correct when the context of the test is clearly delineated, but none of them is the desired answer of measuring the difference in air pressure between the top and base of the building. The point of course is that when we are studying classical physics, the question of how a measurement is supposed to be made is just not germane to the physics being discussed. Why, then, should it become so important in quantum mechanics that the question of measurement become central to the discussion? Why does the mainstream of quantum mechanics insist upon a schizophrenia in which the philosophical interpretation of quantum mechanics has to differ from that of the remainder of physics? In that regard, Wigner's second way demands schizophrenic behavior from physicists.

So it appears that there are reasons to question Bohr's philosophical viewpoint, especially if rational thought leads one to support the idea of Wigner's first way. The issue is not directly a question as to whether Bohr or Einstein and the wave proponents are correct. According to experiment, they both seem to be correct at some level, although we argued in the previous chapter that the concepts of wave function collapse and reality through measurement have significant difficulties with modern measurements. But the ideas of these two camps, plus many other interpretations that differ among themselves, are perhaps best summarized by Feynman [24]:

> Suppose you have two theories, A and B, which look completely different psychologically, with different ideas in them and so on, but that all the consequences that are computed from each are exactly the same, and both agree with experiment. The two theories, although they sound different at the beginning, have all the consequences the same, which is usually easy to prove mathematically by showing that the logic from A and B will always give corresponding consequences. Suppose we have two such theories, how are we going to decide which one is right? There is no way by science, because they both agree with experiment to the same extent. So two theories, although they may have deeply different ideas behind them, may be mathematically identical, and then there is no scientific way to distinguish them.
>
> However, for psychological reasons, in order to guess new theories, these two things may be very far from equivalent, because one gives a man different ideas from the other. By putting the theory in a certain kind of framework you get an idea of what to change. There will be something, for instance, in theory A that talks about something, and you will say "I'll change that idea here." But to find out what the corresponding thing is that you are going to change in B may be very complicated—it may not be a simple idea at all. In other words, although they are identical before they are changed, there are certain ways of changing one which looks natural which will not look natural in the other.

In quantum mechanics, however, we really don't have two separate *theories*. It is true that we have both wave mechanics and matrix mechanics, but their equivalence was shown already by Schrödinger in one of his original papers on quantum mechanics in 1926. So, we have only a single theory with two (or perhaps many) differing formulations. The dramatic difference lies in the philosophical interpretation of the theory. Here, there are many interpretations, one in particular from Niels Bohr and a second which I will call the Einstein–de Broglie–Schrödinger–Bohm interpretation, although it is known that Einstein was not particularly a fan of Bohm.

There are of course others, some of which have been mentioned in previous chapters. All of these others have arisen as attempts to remove the confusing state of our understanding of quantum mechanics (or lack of understanding as Feynman remarked). It is Bohr's interpretation, in which reality arises from the measurement, which is usually referred to as the orthodox theory. Why has Bohr's interpretation, which requires the physicist to have a schizophrenic personality, survived so long? Is it because Bohr's arguments are so intellectually sound that they are unquestionable, or is it because of the determination of the "enlightened cognoscenti" not to admit the error of their ways. It is clear from the above discussion that the first reason is not correct, but it is not clear that the second is correct either. But there are other descriptions which could be applied to the Bohr interpretation.

Disillusionment

The importance of quantum mechanics for atoms began with Bohr's first paper on the subject where he began with Planck's quantization of photons and found the observed experimental results characterized by the Rydberg formula [25]. In this paper, he found at the end of the calculation that the angular momentum was quantized to a value $h/2\pi$. In his second paper, he reversed the calculation and began with this quantized angular momentum [26]. In the third paper, he considered molecules, but begins with the same quantization of the angular momentum of the single electron [27]. For sure, in the later papers he wanted to rid himself of the dependence upon Planck's black body radiation theory, for in this third paper, he writes [27],

> Planck's theory deals with the emission and absorption of radiation from an atomic vibrator of constant frequency, independent of the amount of energy possessed by the system.... The assumption of such vibrators is, however, inconsistent with Rutherford's theory, according to which all the forces between the particles of an atomic system will vary inversely as the square of the distance apart.

But, in these papers, he hasn't spent time on the philosophical implications of the theory and how quantum mechanics should be interpreted.

Needless to say, his theory was criticized and questioned by many different authors. While he began by responding to each, he summarized the various criticisms and his response a couple of years later [28]. Here, we begin to see the formation of the philosophy, as he asserts that the equations are justified by the formulation of Rydberg's parameterization of the experimental observations. He also suggests, in connection with one of the formulas, that "apart from the question of the detailed theoretical interpretation of the formula ... it seems that it may be possible to test the validity of this formula by direct measurements" [28]. While this may seem to be a normal experimental practice, the suggestion here may be regarded as stating that the measurement and confirmation of the formula would justify the theoretical development. In this paper, another observation is that the rapidly developmental style of Bohr's response is appearing, as he answers most criticisms by restating his own derivations with the suggestion that they are the correct form. It is also interesting that from this time he begins to publish in the German literature rather than the British, and doesn't return to responding in the latter until well after the war has concluded. However, when he did return, he published an important paper in which he summarized the current state of the quantum theory and how it fit to a wide range of atoms and experiments [29]. In this latter paper, he embraces the importance of Planck's discoveries and traces it through the subsequent work of many contributors, as has been provided in the earlier chapters of this book. One then questions why he praised Planck in this work, but then ignored him in his own work with Kramers and Slater the same year. Of equal importance, he reviewed the very close relationship between the quantum theory and the results that would be obtained classically. This he describes through the correspondence principle, "according to which the occurrence of transitions between the stationary state accompanied by emission of radiation is traced back to the harmonic components into which the motion of the atom may be resolved and which, according to the classical theory, determine the properties of the radiation to which the motion of the particles gives rise" [29].[b] So, here he seems to be endorsing the close connection between the early (or "old") quantum theory and classical theories. Yet, he would appear to reverse this view within a few years with his dominant demand that measurement is the key element in defining the quantum system.

It is clear that I have described Bohr as a dedicated positivistic in the Machian mold. Whether he was this way at the start or gradually became more of one as time went on is hard to say, and many other authors have held with the view that Bohr was a card carrying positivist. Yet, Whitaker points out that there are many who hold other views [30]. Whitaker describes Folse as suggesting that it is the classical idea of realism, supposedly held by Einstein, that Bohr wishes to renounce. For sure, we find Folse actually saying [31],

> Bohr's association with positivism seems credible in the light of his contribution of a short introductory statement to Vol. 1, No. 1 of the *Encyclopedia of Unified Science*, a project motivated by positivist philosophers of science. But unless we apply the principle of guilt by association, it is clear from all of his writings that Bohr never adopted a positivist model of science.

But this is countered by the observation that [31]

> during the Spring of 1925, Bohr reluctantly agreed for the first time to completely abandon any attempt to describe the atomic system in terms of 'visualizable' or pseudo-mechanical models . . .
>
> Heisenberg saw Bohr's doubts about "visualizable models" as implying that a theoretical representation of the atomic system could and should proceed without attempting a space-time description of the physical conditions actually obtaining within the atom system.
>
> Bohr's concern to represent in his imagination a model of the physical situation which in the observational interaction causes the phenomena predicted by the mathematical formalism did not mean that he regarded such a representation as literally a picture of the atomic system.

---

[b]Let us not overlook the fact that there is a basic difference. Whereas the classical emission would correspond to the energy of the level itself, the quantum emission corresponds to differences in the energy levels.

... Bohr rejected Heisenberg's operational presupposition that nature must imitate a mathematical scheme.

Here, we clearly see that Bohr apparently has adopted the *third way* discussed by Wigner as described above. Finally, Folse [31] tells us that, following the Einstein–Podolsky–Rosen discussions,

> Bohr altered his argumentation by emphasizing that the description of nature is free from ambiguity and metaphysical dogmatism only if it is realized that the observational basis of science depends upon devising a way to describe unambiguously an individual observational interaction between physical systems.
>
> Thus Bohr makes it obvious that such classical terms as "position" and "momentum" are "deprived of all meaning" apart from the context of their application to describe particular observational interactions of phenomenal objects as they appear in specific observational interactions.

This seems to be about as clear a statement of support for positivism as can be found in Bohr's writings.

Whitaker also cites Honner as classifying Bohr as other than a positivist [30], but this arises by stretching some interpretations. Honner tells us [32],

> One formulation of positivism ..., the instrumentalist philosophy of science, reduces science to the ordering of phenomena or observables. In this phenomenalist scheme, science has little or nothing to do with the unseen nature of things. Instrumentalistism is at best very weakly realist. Bohr, given his focus on the observation of quantum processes rather than on quantum objects themselves, is said to belong in such a category.

Hence, by redefining what one calls the most descriptive trait of positivism, we now have a class of *weak realism*, which makes Bohr sound almost as if he agrees with Einstein. As we have noted throughout this tome, nothing could be further from reality. Murdoch addresses this same point [33]:

> Bohr's philosophy of physics is not scientific realism. Is it, then, some weaker form of realism....
>
> Bohr holds that the experimental statements of quantum mechanics, under appropriate conditions, have determinate truth-

values, and hence describe physical reality. . . . he held a non-realist
view of the cognitive status of applied mathematics.

Once more, in an attempt to free Bohr from designation as a
positivist, we find an attempt to redefine various concepts. These
approaches seem to me to appear as various apologies for Bohr's
definitions of reality; attempts to make it seem that he was not so
far from other views of the philosophy of quantum mechanics, views
that clearly held out for a clear idea of reality in the theoretical
foundations, irrespective of any observations.

Having dutifully brought the readers' attention to these attempts
to relieve Bohr of the designation as a positivist, Whitaker then
returns to the major point [30]:

> It may be pointed out that the positivist approach to quantum
> theory would appear to provide fairly straightforward solutions to
> the various difficulties. One could concentrate on the experimental
> results, regarding the only goal as their being expressed in a
> simple mathematical form, and dismissing all deeper speculation
> as meaningless.

Here, Whitaker seems to have perfectly summarized the world view
adopted by Bohr. It states clearly and succinctly the gist of various
arguments that Bohr put forward over the years. His leading acolyte
certainly agreed with this view, when we see that [34]:

> the essential elements of Heisenberg's Copenhagen interpretation
> are complementary descriptions (among which are mutually ex-
> clusive deterministic and space-time ones), a reality that becomes
> actual or definite only upon observation, and a commitment to the
> completeness of such a description.

So, there is little doubt, at least to me, that the Copenhagen–
Göttingen world view is definitively positivist. How, then has this
world view survived through the century or so since quantum
mechanics arrived on the scene? Clearly, it apparently survived the
onslaught of Einstein, de Broglie, Schrödinger, Kennard, and Bohm,
and the acolytes continue to defend it against Gell-Mann, Hartle,

Omnès, and Griffiths. For sure, Bohr defended his philosophy with a rampant pugnaciousness. Other members of the Copenhagen–Göttingen crowd certainly assisted in this task. And, as Gell-Mann has stated [35], Bohr certainly appeared to brainwash his acolytes, perhaps even to the level of creating a cult.

Yet, yet, yet . . . in his later years, some disillusionment must have crept into Bohr's mind. The fact that this was the case certainly is indicated by his final admission, only a couple of years after the Einstein–Podolsky–Rosen paper, [36]:

> On closer consideration, the present formulation of quantum mechanics in spite of its great fruitfulness would yet seem to be no more than a first step . . . we must be prepared for a more comprehensive generalization.

Here, we have Bohr's capitulation, for he is admitting that Einstein and the EPR argument have definite validity—quantum mechanics is, and remains, incomplete. Here, he disowns one of the premier tenets of the positivist case. Bohm's return to causality and determinism, even if only in a probabilistic sense, drives a further stake into the positivist heart. And the modern measurements, discussed in the previous chapter, bring into question the tenets of quantum jumps and collapse of the wave function. With all of this in mind, we might say that the Copenhagen–Göttingen positivist approach has devolved to an empty shell. Realizing this, one may then ask just what the modern acolytes are defending? One may as well ask why more than 900 people followed Jim Jones into suicide in Jonestown, Guyana, apparently in the name of religion? Once a cult is formed, the belief of the followers goes beyond faith, and into fanaticism. That is, it goes beyond just belief in the preaching of the leader, but into a fanatical devotion that leads them to act merely because the leader "says so." How else can you explain physicists who require themselves to be schizophrenic—to have one view of most science and an entirely opposite view of the foundations of quantum mechanics? I suppose that we could in fact say that the Copenhagen–Göttingen view has become a religion, that Bohr followed well known methods for instilling unremitting faith in cult members. This would certainly explain the schizophrenic devotion to the Bohr cause.

In the End . . .

Does it really matter? If one just wants to get on with the computation and find solutions to the equations for a particular situation, the answer is probably "no." But, if we desire to find a deeper understanding into the meaning of quantum mechanics, the answer is decidedly "yes." If we are to summarize the distinction between the positivist crowd and the wave-reality group, it is convenient to use Bohm's depiction of the Einstein–Podolsky–Rosen experiment, as discussed in Chapter 8. The decay within an atom creates two particles that then separate into two opposite directions with opposite momentum. Each of the two particles has a particular spin, and the two also are opposites. In the positivist view, the two spins are completely undetermined prior to a measurement of one being made. Then, the other spin jumps into the opposite state due to what Einstein calls a "spooky action at a distance." Such an action is in complete opposition to the understanding of relativity theory. On the other hand, the realist view is that the two spins are completely determined at the moment the two particles are created, but that we have no knowledge of what those two spins are. We only get to know the values when the measurement is made, but no "spooky action at a distance" is required. The Bohm entangled wave function is now relegated to a statement about our ignorance. For sure, the two particles are entangled and are a single quantum system, but that places no requirement upon whether or when we determine the orientation of the spins, and upon whether the spin orientation occurs at the moment of measurement or at the moment of creation. Einstein, Podolsky, and Rosen go beyond Bohm's simpler model, and discuss the possible wave functions of the two particles. It is from this discussion that they conclude that quantum mechanics is most likely incomplete. Bohr's capitulation, discussed just above, suggests that this conclusion is likely correct. We could paraphrase Bohr's comments above as saying, "Einstein was right."

But Bohr goes further in changing his apparently earlier view on causality in which it is soundly renounced by the Copenhagen–Göttingen group. In this later paper, he falls back to express it [36]:

> The renunciation of the ideal of causality in atomic physics which has been forced upon us is founded logically only on our not being any longer in a position to speak of the autonomous behavior of a physical object, due to the unavoidable interaction between the object and the measuring instruments which in principle cannot be taken into account.

Now, it seems that the causal nature of the quantum system is only ended by the interaction with the measuring system that produces an unavoidable *change* in the system, but this is not that different from measurements in classical systems. And weak measurement theory is designed to avoid these changes in the system, as was discussed in the previous chapter. So, it seems that the question of causality is reduced to whether or not we can describe the measuring interaction sufficiently well to include it within the natural temporal evolution of the quantum system.

For sure, the question of measurements in quantum systems is an old one, and of course dates to the birth of quantum mechanics, but the question of measurements in any system, classical or quantum, is older. The question can be rephrased to: How does the time reversible nature of mechanics, whether Newton's equations (as modified by relativity in that case) or, e.g., the Schrödinger equation, evolve into the time irreversible situation of real life and thermodynamics? Many have searched for this and argued about the so-called arrow of time. This question has probably been discussed since mankind begin to think (and I won't enter any discussion about when that was). Whether classical or quantum, the system is usually described as evolving via the presence of a Hamiltonian for the total energy in the system. Prigogine searched for the form that a decohering interaction should have to induce irreversibility [37], as did Lindblad [38] (and many others). Generally, it may be stated that one needs a loss of coherence in order to connect to the time irreversible world [39], and the form of the Hamiltonian term responsible is often called the Lindblad form [40]. A particularly well-known model, called the GRW model, was proposed by Ghirardi, Rimini, and Weber [41]. This decoherence is certainly key to both the consistent histories approach attributed to Gell-Mann, Griffiths, Hartle, and Omnès and to the decoherence approach of Zurek, both mentioned in Chapter 8. So, it seems clear that Bohr's concern that it is the measurement that upsets causality can be overturned through modern approaches to introducing decoherence into the natural evolution equations for the quantum system. As a result, one more of the tenets of the Copenhagen–Göttingen positivist view has been torn away.

Finally, it seems that the basis of quantum mechanics is clearly described by the Schrödinger equation with its evolution through Kennard and Bohm and its philosophical foundations given by Einstein and de Broglie. It has been extended to handle decoherence and the arrow of time. There is no real reason to resort back to the failed arguments of the positivist crowd. Even matrix mechanics, the form found by Heisenberg, is found within wave mechanics, as expanding the wave function in a set of basis functions, like a Fourier series, leads to the matrix formulation. What Niels Bohr did, with his positivist conspiracy, was exactly as Gell-Mann stated—he set back the understanding of the foundations of quantum mechanics, now by more than a century.

And while the situation seems clear to me, I am sure that there will be arguments, papers, blogs, and webpages that continue to hyperventilate over the mystery of quantum mechanics, seeking only to perpetuate the schizophrenic view of the field. My hope is that the reader now will be prepared to ignore these.

References

1. R. Feynman, *The Character of Physical Law* (The MIT Press, Cambridge, 1965), p. 129.
2. *Ethics in America*, produced by Columbia University Seminars on Media and Society (1989), sponsored by the Annenberg Foundation, lecture 8, http://www.learner.org/resources/series81.html#

3. J. M. Jauch, in *Foundations of Quantum Mechanics, Proceedings of the International School of Physics "Enrico Fermi"* (Academic Press, New York, 1971), pp. 20–55.

4. E. Wigner, *ibid*, pp. 1–19.

5. F. Selleri, *ibid*, pp. 398–411.

6. N. Bohr, H. A. Kramers, and J. C. Slater, *Philos. Mag.*, Ser. 6, **47**, 785 (1924).

7. I. I. Rabi, *Phys. Rev.* **49**, 324 (1936).

8. P. Meystre and M. Sargent III, *Elements of Quantum Optics*, 2nd ed. (Springer-Verlag, Berlin, 1991), pp. 89–92.

9. M. A. Nielsen and I. L. Chuang, *Quantum Computation and Quantum Information* (Cambridge Univ. Press, Cambridge, 2000), Sec. 7.5.

10. I. I. Rabi, in American Institute of Physics Oral Histories: https://www.aip.org/history-programs/niels-bohr-library/oral-histories/24205-2

11. J. von Neumann, *Mathematische Grundlagen der Quantenmechanik* (Springer, Berlin, 1932).

12. A. Shimony, in *Foundations of Quantum Mechanics, Proceedings of the International School of Physics "Enrico Fermi"* (Academic Press, New York, 1971), pp. 470–478.

13. M. Beller, *Osiris* **14**, 252 (1999).

14. W. Heisenberg, *Z. Phys.* **43**, 172 (1927).

15. See, e.g., A. A. Clark, M. H. Devoret, S. M. Girvin, F. Marquardt, and R. J. Schoelkopf, *Rev. Mod. Phys.* **82**, 1155 (2010).

16. E. Schrödinger, *Naturwiss.* **23**, 807, 823, 844 (1935), trans. by J. D. Trimmer, *Proc. Am. Philos. Soc.* **124**, 323 (1980).

17. A. Einstsein, *Phys. Z.* **18**, 121 (1917); trans. by B. L. van der Waerden, in *Sources of Quantum Mechanics* (Dover, Minneola, NY, 1968), pp. 63–78.

18. D. Bohm, *Phys. Rev.* **85**, 180 (1952).

19. E. P. Wigner, in *Foundations of Quantum Mechanics, Proceedings of the International School of Physics "Enrico Fermi"* (Academic Press, New York, 1971), pp. 122–123.

20. A. Schulze-Halberg and J. M. C. Jimenez, *Symmetry* **1**, 115 (2009).

21. D. K. Ferry, *J. Comput. Theor. Nanosci.* **8**, 953 (2011).

22. M.-C. Chang and Q. Niu, *Phys. Rev. B* **53**, 7010 (1996).

23. N. R. Cooper, B. I. Halperin, and I. M. Ruzin, *Phys. Rev. B* **55**, 2344 (1997).

24. R. P. Feynman, *op cit.*, p. 168.

25. N. Bohr, *Philos. Mag.*, Ser. 6, **26**, 1 (1913).

26. N. Bohr, *Philos. Mag.*, Ser. 6, **26**, 476 (1913).

27. N. Bohr, *Philos. Mag.*, Ser. 6, **26**, 857 (1913).

28. N. Bohr, *Philos. Mag.*, Ser. 6, **30**, 394 (1915).

29. N. Bohr, *Nature* **112** Suppl., 29 (1923).

30. A. Whitaker, *Einstein, Bohr, and the Quantum Dilemma* (Cambridge Univ. Press, Cambridge, 1996), p. 167.

31. H. J. Folse, *The Philosophy of Niels Bohr* (North-Holland, Amsterdam, 1985).

32. J. Honner, *The Description of Nature: Niels Bohr and the Philosophy of Quantum Physics* (Clarendon Press, Oxford, 1987).

33. D. Murdoch, *Niels Bohr's Philosophy of Physics* (Cambridge Univ. Press, Cambridge, 1987).

34. J. T. Cushing, *Quantum Mechanics: Historical Contingency and the Copenhagen Hegemony* (Univ. Chicago Press, Chicago, 1994).

35. M. Gell-Mann, in *The Nature of the Physical Universe: 1976 Nobel Conference*, ed. by D. Huff and O. Prewett (Wiley-Interscience, New York, 1979), p. 29.

36. N. Bohr, *Philos. Sci.* **4**, 289 (1937).

37. See, e.g., I. Prigogine, *From Being to Becoming* (W. H. Freeman, San Francisco, 1980).

38. G. Lindblad, *Non-Equilibrium Entropy and Irreversibility* (D. Reidl, Dordrecht, 1983).

39. See, e.g., L. S. Schulman, *Time's Arrow and Quantum Measurement* (Cambridge Univ. Press, Cambridge, 1997).

40. A. M. Ozorio de Almeida, in *Entanglement and Decoherence*, ed. by A. Buchleitner, C. Viviescas, and M. Tiersch (Springer, Berlin, 2009).

41. G. C. Ghirardi, A. Rimini, and T. Weber, *Phys. Rev. D* **34**, 470 (1986).

Index